Breakthrough!

BREAKTHROUGH!

CANADA'S GREATEST INVENTIONS AND INNOVATIONS

JOHN MELADY

DUNDURN
TORONTO

Editor: Shannon Whibbs
Design: Courtney Horner
Printer: Webcom

Library and Archives Canada Cataloguing in Publication

Melady, John
    Breakthrough! : Canada's greatest inventions and innovations / John Melady.

Includes bibliographical references.
Issued also in electronic formats.
ISBN 978-1-4597-0852-5

    1. Inventions--Canada. 2. Technological innovations--Canada.
I. Title.

T23.A1M44 2013   609.71   C2013-900783-0

1   2   3   4   5   17   16   15   14   13

We acknowledge the support of the **Canada Council for the Arts** and the **Ontario Arts Council** for our publishing program. We also acknowledge the financial support of the **Government of Canada** through the **Canada Book Fund** and **Livres Canada Books**, and the **Government of Ontario** through the **Ontario Book Publishing Tax Credit** and the **Ontario Media Development Corporation**.

Printed and bound in Canada.

Visit us at
Dundurn.com
Definingcanada.ca
@dundurnpress
Facebook.com/dundurnpress

Dundurn	Gazelle Book Services Limited	Dundurn
3 Church Street, Suite 500	White Cross Mills	2250 Military Road
Toronto, Ontario, Canada	High Town, Lancaster, England	Tonawanda, NY
M5E 1M2	LA1 4XS	U.S.A. 14150

This book is for my grandchildren:
Hillary, Hannah, Liam, Conlan, Keegan, Mackenzie, Meg, and Riley.
With my love to all of you.

CONTENTS

PREFACE

Canadians are behind praiseworthy products, fascinating ideas, and machines with global appeal. We have invented with little fanfare, financing, or expectation. Yet our creations are important. The greatest is insulin; millions use it or die. Then there is the light bulb that illuminates the world. And a Canadian perfected a kind of wheat that when grown in our prairie provinces and elsewhere helped to feed human beings in countries around the globe.

And we share our creations: Bell conceived the idea of the telephone in Canada; then sold it in the United States. Naismith invented basketball and Americans play it. The Canadarm is Canadian, and lifts in outer space. As well, the men and women who are, or have gone into space, along with fighter pilots everywhere wear pressurized suits that were invented and initially designed here. The wirephoto was the brainchild of a man from Winnipeg named William Stephenson. He tried to sell it in Canada, but found that no one was interested. When he took it to Britain, it was an immediate success. A technician in Ottawa became alarmed when he realized planes were crashing in remote locations and becoming lost forever. The device he pioneered was able to tell searchers where to look. In time, the inventor's idea became the Black Box that is found on planes today, the world over.

Some of our inventions are physically small: the BlackBerry, the paint roller; the Robertson screwdriver. Others are larger: the jetliner, the snow blower, the snowmobile. The pacemaker saves hearts; the goalie mask saves face, and the Festival saves Stratford. We have national identifiers: a donut shop named Tim's and a maple leaf flag that flies over a nation that is widely envied.

ACKNOWLEDGEMENTS

Every author of every book has help in bringing the volume to the reader. This book was no exception. I was helped by many individuals, and it is impossible to thank all of you. One of the reasons was because I did not know your name, or because you offered assistance, and then you were gone. This often occurred when I was asking for directions in my search for individuals to interview, or particular files I hoped to locate. Those of you who are librarians were of particular assistance and are a credit to your profession. Rarely are you thanked as often as you should be, so please accept this token of appreciation from someone who is truly grateful.

As well, I want to extend my sincere thanks to Paul Moran, Jan Hawley, Susan Hundertmark, the wonderful staff at the Seaforth Public Library, Peter Bob McDowell, Jim Brown, Paul Vick, Larry Zaleski, Shauna Duff, the helpful individuals at the Tim Hortons head office who did not want to be specifically mentioned, Jean-Pierre Arsenault at the Canadian Space Agency, Kelly Masse at the Hockey Hall of Fame, the staff of the Sir Frederick Banting Museum in London; Brian Wood, curator at the Bell Homestead in Brantford; the individuals who assisted me at the National Archives in Ottawa; those at the Bombardier Museum in Valcourt; and the personnel at the Robertson Archive in Milton. A special thanks goes to Nels Banting for his insights about his famous uncle.

As always, I must thank several individuals at Dundurn, in particular Michael Carroll, Shannon Whibbs, Cheryl Hawley, Courtney Horner, and Jennifer Scott. Thanks as always to my wife Mary for the encouragement and assistance.

1

FREDERICK BANTING
Lifesaver for Millions

There are millions of people alive today because of one determined Canadian. His name was Frederick Grant Banting, a young physician with a struggling practice in London, Ontario, in 1920. The doctor was a veteran of the First World War, and had been wounded by shrapnel during the carnage at Cambrai in September 1918. He returned to Canada with a Military Cross, and an uncertain future.

With money borrowed from his father, he bought a two-storey yellow-brick house in a residential area a short distance from downtown London. The man who sold him the property lived there until a house he was building was completed. In the interim, Doctor Banting had the use of the front parlour where he would see anyone who was ill, and an upstairs bedroom where he slept. He did not need more space because his patients were few. He opened his practice on July 1, 1920, and for that first month, his total receipts amounted to only four dollars. Life was bleak indeed.

Banting was a rough-hewn country boy who was raised on a farm just east of Alliston, Ontario. He had been born in a downstairs bedroom of the family farmhouse, and was the youngest of five children. His father worked the land, raised crops and cattle, and provided for his family in an adequate, though not extravagant, fashion. While he was growing up, Fred did all the jobs every farm boy did. He knew what work was, whether feeding livestock, cutting grain, or forking hay.

Library and Archives Canada PA-123481.

Doctor Frederick Banting in a photograph taken in Toronto *circa* 1920–25. This was at the beginning of his remarkable rise to fame, and not long after the discovery of the wonder drug that has saved the lives of millions of people.

The Banting house in London, Ontario, where Doctor Frederick Banting lived and practised medicine when he formulated a cure for diabetes.

He loved animals: the heavy horses that pulled wagons and ploughs, the cows that had to be milked morning and night, and the succession of dogs that were family pets. He was particularly fond of the dogs.

Growing up in the country was appealing, and perhaps owning his own farm later on could have been his future. However, Fred wanted to do something else, and get an education beyond what local schools could offer. He was not a particularly good student, but thoughts of exploring other fields intrigued him, in spite of his decidedly mediocre learning record. With the blessing of his parents, he left Alliston to study at the University of Toronto. He enrolled in the medical school there, and in time graduated as a doctor. Around this time, he also became engaged to be married, but because the First World War was raging in Europe, he volunteered for duty, donned a uniform, and soon saw first-hand the waste of humanity that conflict wrought.

After the cessation of hostilities, he returned to Canada and not long afterwards opened his London office. He went there, in part, because his fiancée, whose name was Edith Roach, was teaching school in nearby Ingersoll. Her salary at the time was a relative pittance, but it was still more than Fred made — and this bothered him. To help sustain himself until his practice picked up, he began working part-time at the medical school at the University of Western Ontario. Here, he was paid $2.00 an hour.

It was within this context that Banting became increasingly worried about where his life was going. His patients remained few; his university job was inadequate, and Edith Roach was becoming ever more demanding. She wanted to marry, but Banting felt he was not ready to do so, and in any case felt that a husband should support his wife. As yet, he could not afford to do so. However, he continued to strive. He sometimes lectured at the university and as part of his job, was occasionally asked to speak on medical subjects that he did not know well — or just partly understood. This was the case when he found himself scheduled to give a talk to medical students on October 31, 1920.

The topic that day was to be about the working and function of the pancreas, and he intended to describe what happened to the body when the pancreatic duct was blocked by gallstones. Even though he researched the topic as well as he could in the time he had, he still felt that he barely understood it. However, he became convinced that something occurred during this blockage that might have some connection to the onset of diabetes, a terrible disease that left those with it "listless and suffering from progressive weight loss. The only treatment [at the time] was a strict diet that was generally incapable of sustaining a healthy body. Thus diabetes was often a death sentence for its victims."[1]

Banting read and reread a scientific journal article about the matter just prior to going to bed the night before his lecture. Thus, the pancreas and its function was on his mind as he tried to get to sleep. He tossed and turned restlessly for an hour, then longer, but sleep would not come. Finally, he got up around 2 a.m., picked up a pen beside his bed, and jotted down a few words on a sheet of paper. These words had to do with an idea he had concerning a possible treatment for diabetes, the killer that had been around since the ancient Greeks wrote of it. He was well aware of the fact that even though there had been lots of reported cures for the illness, none of them had worked.

As it turned out, the few words he wrote provided a core idea that would soon lead to a cure for diabetes, and would change Doctor Banting's world forever. Years later, he explained to an audience the essence of the idea: "Ligate pancreatic ducts of dog. Wait six to eight weeks for degeneration. Remove the residue and extract."[2] Within the thought was the key that would lead to the wonder drug that would save the lives of sufferers around the globe.

When morning came, Banting made his way to the Western campus, gave his lecture as scheduled, but continued to ponder his overnight musing. From what we can understand today, he really was convinced that he was on to something, but before he could do anything with the idea, he knew he had to prove to others that he could be right. That would not be easy.

The young doctor turned to a couple of medical professors he knew at Western, and asked their advice. Both listened to his suggestion, and perhaps to humour him; perhaps because they may have felt that he was right, suggested a further step. That would involve consulting an acknowledged expert on the subject of diabetes — a doctor from Scotland — who happened to be teaching at the University of Toronto at the time. The man's name was J.J.R. Macleod. Banting decided to go and see him.

The meeting did not go well.

Macleod was a haughty, opinionated, somewhat pompous specialist in his field. He appeared to have little regard for the man who sought his direction, and was initially rather dismissive. Even as Banting stood before him, explaining his idea, Macleod turned to read correspondence on his desk. Banting was not impressed. Years later, in another context, he referred to the short-statured Macleod as a "goddamned little bugger."[3] Yet, the meeting was the foundation for what followed.

Banting asked Macleod for some laboratory space in the medical facilities at the Toronto university and a small number of dogs that could be used for research. Even though Macleod demurred initially, he did give the idea more thought after his visitor had gone. A disappointed Banting returned to London, but refused to give up. A short while later, he wrote to Doctor Macleod and repeated his request. This time, the eminent professor hinted at a way out, so Banting went back to see him.

Macleod was leaving Toronto for the summer, and would be spending the time in his beloved Scotland. He rather grudgingly offered Doctor Banting a small, decrepit, rarely used storeroom in a medical building, and said a few dogs could be made available as needed. The animals were already at the university. In addition, Banting would be able to hire help for the work he was about to do. To that end, and not long afterwards, a twenty-two-year-old medical student named Charles Best became Banting's assistant. Then, despite occasional differences of opinion between them, Banting and Best were the two-man team who did the essential legwork in the search for a diabetes cure. Later on, a biochemist named J.B. Collip joined them. And despite the fact that he was far from Toronto in those critical first weeks, Doctor Macleod was technically in charge of the project itself. His role would later be subject to both scrutiny and dispute. However, he did assist in some of the groundwork in showing Fred Banting how to operate on research dogs. In time, Macleod also performed a more public role when the results of the research showed promise.

The summer of 1921 was extremely hot and humid in Toronto, and air-conditioning as we know it today was far in the future. Yet, in spite of working conditions that bordered on the impossible, Banting and Best went about their study of what would happen to the dogs they operated on in order to "permit the isolation of the anti-diabetic component of the pancreas."[4] The work went on for days and days, and sometimes seemed to be utterly futile. For example, all of the first dogs died in the search for the answers that were sought. This broke Banting's heart because to him, all dogs were special. In one instance, a diabetic animal that had been kept alive using the extract that he and Best had isolated became severely ill, and then passed away. "When that dog died," he wrote later, "I wanted to be alone for the tears would fall despite anything I would do."[5] And yet, he and Best worked on. They had some success, but too often failure. Ultimately, however, they determined that the extract from the pancreas was the key they were seeking. They continued the animal experiments.

Over time, diabetic dogs that were injected with the extract regained their health. Some that had been hours from death recovered and became affectionate and playful. The two men rejoiced in what they had accomplished, and "on August 9, Banting wrote to Macleod in Scotland, announcing that he and Best had an extract that improved the clinical condition of diabetic dogs. Banting wondered if the extract would work on humans."[6]

The Scottish physician's initial reaction was to play down the idea of any positive conclusion. He suggested several things that could have come into play. He questioned the results he was shown. He faulted the research techniques of the two men who had laboured all summer on the project. To a degree he was playing a devil's advocate role, which was probably needed at the time. However, Banting was annoyed and showed it. He and Macleod argued often; neither man liked the other, and in retrospect, it is almost a wonder that the study continued at all. Yet, neither Banting nor Best wanted to give up. They continued their research and gradually were able to obtain, purify, and then use the pancreatic extract first of dogs, then of other animals in a Toronto slaughterhouse.

Finally, they had enough success treating dogs that they felt the time had come when they should try injecting humans. At this point, they called the extract "isletin," but the name was later changed to "insulin" as we know it today. It is believed that the first human injected was Doctor Banting himself. When he suffered no side effects, he was sure the time had come to test someone else. The person selected was fourteen-year-old boy. His name was Leonard Thompson.

The teenager was then a patient at Toronto General Hospital. He weighed only sixty-six pounds and was a severe diabetic. His life expectancy was almost non-existent, and the doctors treating him had pretty much given up all hope of his survival. With nothing to lose, the Banting/Best extract was injected into him. Initially, it was not a success, mainly because the substance was not completely pure. A better sample was prepared and used, and almost immediately, the sick boy improved. Insulin worked! However, myriad problems remained — some of which dogged the researchers for weeks to come, and often seemed insurmountable.

The major stumbling block lay in the shortage and quality of the extract. It was difficult to make, and even harder to purify. Doctor Collip looked after that aspect of the work, and at times had much success. However, there were often critical junctures where he had great difficulty developing the exact formula for the drug. At times of shortage, the pressure on Banting and his colleagues was acute. News of the apparent cure had leaked to the press, and any snippet of positive coverage was seized upon by diabetics, their families, and their doctors. Medical clinics in Canada, the United States, and Europe were besieged with diabetic patients who were convinced they would die unless they got immediate help. And unfortunately, many did die because there was just not enough insulin to go around.

And during this time, when the pressure to produce was at its height, Doctor Macleod delivered papers about the work of Banting, Best, and Collip at important medical conferences. In doing so, he often used the personal pronoun "we" when he described the early success of the research. When the tenor of Macleod's remarks was conveyed to Banting, he was both hurt and furious — and convinced in his own mind that Macleod was trying to claim personal credit for the discovery. After all, Macleod was an internationally famous professor, and the renowned expert on the subject of diabetes. By comparison, Banting was just a poor country doctor who had little following, not much expertise, and almost no understanding of the politics involved in major medical circles. On a scale of achievement, Banting felt inadequate, and that led him to confront the Scotsman. There was a raging argument, and while the two did not come to blows, they were essentially estranged from then on. They never completely reconciled.

The continuing problem lay in securing enough insulin to treat the patients who needed it. To Banting, this was particularly heartbreaking. Extremely ill diabetics came to him; hundreds more wrote, detailing their sufferings, and begging for this wondrous drug; doctors in Toronto and elsewhere approached him on behalf of their patients. And the fact that so many of the ill were poor and with little hope bothered Doctor Banting to no end. He knew that turning them away meant sentencing most to an imminent death.

Fortunately, another "miracle" of sorts came into play about this time. Initially, a Toronto company called Connaught Laboratories produced the drug. However, when the demand for insulin swelled beyond their capacity, an American firm called Eli Lilly, based in Indianapolis Indiana was licensed to make the product. This was because senior officials with Eli Lilly had shown early interest in Banting's discovery, and were convinced of its effectiveness. In turn, doctors and others at the University of Toronto knew of and respected the Eli Lilly brand.

In the months that followed after Fred Banting's fortuitous night-time jotting-down of his idea, his life had become extremely exigent. He had closed his medical practice in London, then sold his house there. His relationship with Edith Roach had ended, and he had moved to Toronto, where he intended to stay. With the discovery of insulin, his days as an unknown medical man were over for good. He was besieged on every side by those who wanted to place him on a pedestal, something he had never desired. He often felt that he had to escape from the hectic existence he experienced on a day-to-day basis. Fortunately, he found a way out, but a much-too-fleeting one.

Back in his high school days, Banting liked art, and to fill in the long hours when he was waiting for patients in London, he started to draw, and then to paint. Initially, he had no idea whether the pictures he created were any good, but in making them he was able to transport himself into a place where he was comfortable and at peace. Subsequently, meeting Group of Seven painter A.Y. Jackson enhanced this love of art. The two men became good friends, and when they could, embarked on painting forays to various locations.

Doctor Frederick Banting's photo on the cover of the August 27, 1923, issue of *Time* is proud-ly displayed at the Banting house in London, Ontario.

They went to the south shore of the St. Lawrence, to places in northern Ontario, and to the western Arctic together. Banting always loved these trips, and longed for more of them. On each, he learned from Jackson, and often the work they produced had a similarity in format and execution. Undoubtedly, "both were men of keen intellect and individualism and shared a love of adventure. Above all, they were passionate about Canada and wanted to express this passion on canvas."[7] Today, many regard Banting, the artist, in a special light. Toronto fine art dealer David Loch says of the man: "Banting wasn't just this famous guy who painted. He was a skilled artist in his own right."[8] Later on, during the Second World War, and just after the two men had returned from a painting expedition, Banting confided in Jackson: "When the war is over, I'm through being a doctor and I'm going to paint for the rest of my life."[9] This was in 1941, and sadly, it was the last time the two were ever together. However, it did illustrate how frustrated Banting had become with the circumstances of his life at that time, and with the crushing fame that had been thrust upon him. He continued to paint whenever he could, and while he did not leave a lot of pieces for posterity, his art today demands ever-escalating returns. For example, $76,050 was the price raised by the Heffel auction house for a 25-x-32-inch Banting oil that sold on November 19, 2008. It has been said that such a figure would have astonished the man who painted it.

Yet it was the fame he never wanted that bothered Banting most of all. He and Macleod won the Nobel Prize for Medicine at Stockholm in 1923, and an aspect of that honour rankled most of all — because Macleod also got it. Banting said that his closest associate Charles Best should have received the prize, and backed up his assertion by splitting the monetary component with Best. Somewhat sheepishly it is said; Macleod then shared his money with Doctor Collip. Sometime later, Nobel Prize Foundation members admitted that Macleod should never have been given the prize. "He had taken no active part in the work, and in fact had been away when the decisive experiment was made,"[10] they stated in 1962, thirty-nine years after they made one of their greatest mistakes.

As might be expected however, the Nobel Prize enhanced the reputation of the University of Toronto. Grateful officials there named a

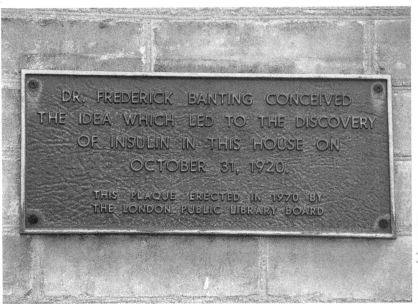

Photo courtesy of the author.

Outdoor wall plaque at the house in London, Ontario, where Doctor Frederick Banting was living when he conceived the idea of a cure for diabetes.

medical facility after Doctor Banting, and the house where he came up with the idea for the diabetic cure in London, Ontario is now a small museum and a National Historic Site. Hundreds of people from all over the world visit the place every year. No record is kept of how many of them are diabetic.

The country doctor who shunned fame was knighted by King George V in 1934, and Sir Frederick Banting remained in the public eye for the rest of his life. Tragically, he perished in a plane crash near Musgrave Harbour, Newfoundland, on February 21, 1941. He was forty-nine at the time.

2

ALEXANDER GRAHAM BELL
Pioneer Communicator

The man who is often called Canada's greatest inventor was technically not a Canadian at all. Nevertheless, Alexander Graham Bell, the practical dreamer who gave us the telephone has to be considered in any account of inventions and this country. By birth and upbringing, Bell was a Scot. He was twenty-three years old when he came to Canada with his parents, and then lived there off and on for the rest of his life. He also died there, and he and his wife Mable are buried on Canadian soil.

Both Bell and the former Mabel Hubbard were Americans; she by birth; he by choice — a choice necessitated primarily for commercial reasons. He felt his chances of patenting and marketing the telephone would be better if he had American citizenship, and in all probability, he was right. At the time, most of the money, the daring, and the opportunities were in the United States. In 1874, when he first came up with the idea for the invention, Canada was a much more conservative country than America. There were not a lot of wealthy backers, and Bell realized that. Possibly, if he had had greater financial support, he might have remained in Canada. He left southern Ontario, where the family had settled, simply because he needed money, and was unable to find it in this country. Throughout his lifetime, because his mother was deaf, his dream was to teach the deaf, and to make ends meet, he took a job in that field in Boston, Massachusetts.

But why did Bell come to Canada originally?

The Bell homestead in Brantford, Ontario, where Alexander Graham Bell conceived the idea of the telephone.

The short answer is: for health reasons. He was born in Edinburgh in 1847, the second of three sons. His father Melville was a speech teacher in the Scottish capital, and a successful one. He and his wife Eliza doted on their boys, but the affection was intermingled with heartbreak. Both the oldest and the youngest died of tuberculosis, and in time Alexander exhibited signs of contracting the disease as well. It was at that point that the family decided to escape the pollution of their city and move to Canada. Alexander's father was sure the air was purer there and it would cure the son's medical condition. As it turned out, he was right. Almost from the time the family had settled in "Melville House," as they called it, near Brantford, Alexander's health improved. He needed less rest, and gradually turned his thoughts from his own welfare to his dreams of improving the condition of those who were unable to hear. To that end, his father found Alexander, or Alex, a teaching position in Boston. The young man went there with mixed emotions, but felt it was time he was out on his own.

And even though Alex loved teaching, and by all accounts was very good at it, he also had a life-long interest in all kinds of other things. This was true when he was a boy in Scotland; it was just as true in Boston, Brantford, and Baddeck, Nova Scotia, where he lived for the last thirty-six years of his life. And if there is an obvious criticism of the man, it was his impetuosity: he would often be working at one thing and get an idea for something else. His fertile mind would jump to the new, and in no time, he would turn his entire focus to it. One such matter that intrigued him was the idea of somehow sending the sound of the human voice along a wire, from one person to another; from one location to another. If he could figure out how to do this, he reasoned, the idea would benefit society.

By the time Alex moved to Boston to teach, electricity existed, and refinements for its use had been invented. One of those was the discovery that electrical impulses could be sent along a wire. A man named Samuel Morse built on that idea when he came up with a code for transmitting messages; a code that was named for him. A bit later, the first telegrams were being sent in the United States, then across North America and elsewhere.

Alexander Graham Bell knew of the much-heralded advances in telegraphy, of course, and he started using his every spare moment to try to figure out some way of transmitting voice instead of code. He taught the deaf during the day, and then stayed up half the night doing experiments of various sorts in order to achieve his theory. Unfortunately, he worked to the point of becoming ill, and found that he had to return home whenever possible in order to rest and regain his strength. These visits with his parents ultimately meant that he was in Canada for at least a part of each summer. The time was not only regenerative; it also meant he could confer and discuss with his father each new idea he had come up with during the winter months. Then, in a workshop and on the grounds of Melville House at Brantford, he dreamed of what could be. That was where he was when he developed the concept for the telephone. That one idea changed his life.

But the time in the United States was always hectic, and Bell did not waste it. For the first while in Boston, his teaching took precedence over everything else due to financial necessity. He had little money, and he had to feed and clothe himself. He dressed poorly, always seemed unconcerned about his appearance, and lived in ramshackle accommodations. However, during his off hours, he was continually trying to invent things. But then, another interest in his life came about by chance. A young, seventeen-year-old girl who enrolled in his speech class became the centre of his world.

Mabel Hubbard had suffered from scarlet fever when she was a child, and as a result, could neither speak nor hear. She was a good student, intelligent and inquisitive, and her teacher fell in love with her, despite the fact that he was twenty years older. Gradually, the attraction became mutual, and after a lengthy, if puritanical courtship, the two decided to marry. Soon, Mabel Bell realized that while her new husband was a wonderful teacher and an inventive genius, he was no financier. For that reason, she had to look after their finances. In fact, though his invention of the telephone would ultimately make him rich, money was never uppermost in Bell's mind. Mabel kept the family books because her husband had no interest in doing so at all.

But he did have lots of other interests, and these were not limited to his inventive nature alone. He was a natural storyteller, and was adept at entertaining in social settings. His admirers would sit transfixed as he related anecdotes about his youth and formative years in Edinburgh. His mother had taught him how to play the piano, and he would often play for guests. By all accounts, those who heard him loved the performances. Then, to further entertain his listeners, he would tell stories relating to Scottish lore, sometimes adapting different voices for different characters of the past. Bell also loved to take long walks in wooded areas, along the seashore, or anywhere else where he could be alone with nature and his thoughts. He also loved books that had "information on sound — books on acoustics, the tuning of instruments, the human ear. He was particularly interested in how the ear worked …"[1] This interest was of vital importance to him when it came to the concept of the telephone.

Bell was in Brantford when he envisioned his greatest invention. There, on the attractive grounds at Melville House, overlooking the Grand River is a secluded location behind the home. When he visited his parents during the summers, Bell often spent hours in this quiet area, alone with his thoughts. He called it his "dreaming place," and he was

Image courtesy of the author.

The Bell homestead in Brantford, Ontario, where Alexander Graham Bell conceived the idea of the telephone.

there on an afternoon in July 1874 when he came up with the idea for the telephone. Much later, in a speech he gave in Brantford, he declared, "I've often been asked if Brantford is the home of the telephone. This I will say: the telephone was invented here."[2] That single statement adds a clarity to the often-argued question as to the location where the telephone originated. Without a doubt, Bell's remark settles the matter.

However, now that he had the idea in his head, Alexander had to see if he could somehow make the concept a reality. That took time, experimentation, and lots of patience. And even before he had a working model for any kind of telephone, he took the advice of advisors in Boston and hurriedly sought a patent for his invention. Once that was granted, he was pressed to built the thing and make it work — quickly. This task was highly stressful for the man. He had built things in the past, or tried to build them, and found doing so to be almost beyond him. Fortunately, he had the presence of mind to admit this inadequacy, and he went looking for an assistant who could make whatever might be needed. The assistant was a pleasant, industrious young man named Thomas Watson who worked at an electrical shop in Boston.

Watson was twenty-two when he went to work with Bell. The young man was mechanically inclined, and over time, would prove it again and again. Bell would get an idea for some kind of machine, sketch out the general plans for the thing, and ask Watson to make it. Invariably, he was able to do so. In fact, Bell had come to rely on Watson's genius for some time, and was unbelievably fortunate to have his assistant with him when he had to make a working telephone on such short notice. Together, the two were a team; a team as close as any in the history of invention. What one conceived; the other executed. Their workplace was a large attic in the house in Boston where Bell lived.

At the time, Alexander still held his teaching job, so the race to build a working telephone took place during the night. There was much trial and error in everything the two men did, and batteries, wires, and assorted bits and pieces of materials were strewn everywhere. Watson made what they called a "receiver" for Bell; then a new transmitter. These were then linked by a wire. At the transmitter end, the wire touched an acid bath, a necessary substance for the transmission of voice, if it came to that.

Finally, late in the evening of March 10, 1876, the two men were as prepared as they ever would be to see if the telephone experiment would succeed. Together, they propped the receiver up on a piece of furniture in one room, and then placed the transmitter in another. Bell sat in front of the transmitter while Watson walked down a hall to the second room, and got ready to position himself at the receiver. By all accounts, he was scarcely there when he heard Bell's voice coming out of the receiver. Alexander had rather clumsily spilled some acid on his pants, and needed immediate assistance.

"Mr. Watson, come here, I want you!" he called.

Watson bounded to his feet and ran to Bell's room, scarcely comprehending what had just happened. Much later, the young man recalled his actions: "I rushed down the hall into his room and found the acid of a battery over his clothes. He forgot the accident in his joy over the success of the new transmitter when I told him how plainly I heard his words."[3] The two men looked at each other, both momentarily awestruck by what had transpired. Then they started laughing, cheering, and slapping each other on the back. In their excitement, Bell's carelessness with the acid was forgotten and the two "spent the whole night talking on the tele-

Alexander Graham Bell, the distinguished inventor who gave us the telephone. This photo was taken between 1914 and 1919, several years after his great invention was in widespread use.

phone."[4] The calls were the first in history, anywhere, at any time. Bell was in one room, Watson in the other. They switched places several times, talked of whatever came to mind, and fine-tuned the primitive apparatus they used. Bell's original idea, from the "dreaming place" overlooking the Grand River at Brantford, Ontario, had become reality. By the time the first streaks of morning light bathed Boston; two determined young men had changed human communications.

But Bell's quest was far from over. As word began to leak out about the marvellous invention, his detractors became more strident. Some people refused to believe what had happened; many others wanted to cash in on the discovery. Almost from the time that first call was made, others claimed they had made the discovery first. For years afterwards, Bell and a handful of backers in his fledgling telephone company had to fight a succession of lawsuits relative to the phone. He always found doing so to be a time-consuming, frustrating nuisance. However, he forged ahead and won his case every time. Doing so was necessary, despite the amount of money that had to be allotted to defending the case.

Then, in the early days after he got his revolutionary machine working, he had to spend time demonstrating it to all kinds of organizations. He did this in several locations in New England in particular. Bell and his early supporters felt these sessions were necessary if the telephone was ever to become accepted. People had to know about it. On the other hand, there were always those who felt it would never succeed, and they would have nothing to do with investing in it. There were many of these, both known and unknown. One of those most prominent in the former group was the writer Samuel Clements, or Mark Twain. He had the chance to buy into Bell's company, and indeed probably would have been given preference in obtaining whatever shares he wanted. He declined.

After a period of what today might be called "in house" demonstrations of the invention, Alexander decided to bring it home to Brantford in order to see how the thing would perform if calls were made from one building to another, or perhaps even between two locations that were far apart. He did this, starting in the summer of 1876, one year after he and Thomas Watson made that first call.

There was considerable excitement in Brantford when Bell stepped from the train that brought him from Boston. By then, news of his amazing invention was already widely known in Ontario, and the locals wanted to see how it worked. Bell did not disappoint, although he was initially hesitant about making the first call over any distance. This had never been done before, and he wanted to try it at home, for his parents, and for those who always believed in him. For that reason, shortly after

Photo courtesy of the author.

Model of the first telephone on display at the Bell homestead and museum in Brantford, Ontario.

he arrived, he placed a transmitter in Melville House and a receiver in an outbuilding a short distance behind it. Then, when that first call from one structure to another worked, he knew he had to see if the calling distance could be increased. That was why, on the morning of August 3, he found himself in a telegraph office in the village of Mount Pleasant, some five miles from Brantford. A small telegraph company there had wires strung between the two municipalities and Bell prevailed upon the owners to let him use their hook-up to test his invention. They were only too happy to do so.

While Bell remained in Mount Pleasant, his uncle David took up a position in the Brantford office, and at the appointed time, spoke into the transmitter there. Bell was immensely relieved when he heard his uncle's familiar voice, but a great deal of static interfered with the clarity. However, when a more powerful battery was hooked up at the Brantford end, the static died out. And, in a dramatic touch to the whole affair, the words "to be or not to be," from Shakespeare's *Hamlet* were the ones Bell heard over the wire. They immediately became part of the folklore that in later years would surround the making of that first long-distance call anywhere.

Following this important development, this new method of communication quickly became popular, and within a couple of years over two thousand telephones were in use. In no time, that number increased dramatically. Bell and his partners became rich, and the inventor was able to pursue other endeavours. He helped found the National Geographic Society, and then he went on to invent other things, in aviation, marine engineering, medical science, and even animal husbandry. Today, people from all over the world flock to his graveside in Baddeck, Nova Scotia, his final resting place after a life that was certainly full.

3

MIKE LAZARIDIS
Genius Ahead of His Time

As seen from the Golden Horn, its natural harbour, Istanbul, Turkey, appears to be one of the most exciting cities in the entire world. Here, where Europe and Asia meet, the great architectural structures on the hills before you are overpowering in their beauty and magnificence. There is the famous Blue Mosque, or more correctly, the Sultan Ahmed Mosque, the Hagia Sophia museum, one of the world's historical landmarks, the Galata Bridge, with its restaurants below the traffic lanes, and farther away, the Grand Bazaar with its colours and crowds, and the spirited haggling for everything from exquisite jewellery to junk. It was undoubtedly a remarkable place to be born, but for many Greek Christians in this Turkish town in the mid 1950s and after, it was not a place to remain.

Istanbul was and is a large city. There are over 13 million souls there now. It is also an old town, with roots going back twenty-six centuries when it was called Byzantium. But in 1955 in particular, it saw much cultural and religious upheaval, and Greek Christians living there were ostracized and targeted by certain factions of the Turkish majority. For that reason, frightened Greeks began to leave the place, and in time, put down roots elsewhere; sometimes for good, sometimes temporarily, until they reached a country of refuge or until the turmoil in their homeland passed. One man who elected to leave and not return was a merchant named Nick Lazaridis. He had a successful business in his home city, but left it with his wife and young son. The boy's name was

Mihal, or Mike, as he would later be called. His mother was Dorothy.

The Lazaridis family left Istanbul in 1964, spent a couple of years in Germany, but then sailed across the Atlantic to Canada. They arrived in Windsor, Ontario, in 1966, when young Mike was five years old. From that time onwards, Canada was home for the family. And as is the case with many, and perhaps most new arrivals to this country, life was initially far from easy. However, neither Nick nor Dorothy was afraid of work. In time, they succeeded in their adopted homeland and never looked back.

Nick Lazaridis found employment with a couple of companies, but eventually gravitated to Chrysler, the auto company. Dorothy had her own sewing shop. In time, these new Canadians bought a house and put down permanent roots in Windsor. This may not have been Istanbul, but it was home. There was no fear of being ostracized because you were not Muslim. Nor was there fear of having your livelihood or business restricted because of local laws that were discriminatory. Canada was a new land, and the cultural mix of recent arrivals made it even better. Here, your life was what you made of it, and success was something everyone could aspire to.

Mike, and soon a younger sister, Cleopatra, went to local schools, and both were achievers. Always, Mike was a popular youngster who was well liked and not surprisingly, made friends easily. One of these was a boy he met in grade six named Doug Fregin. The two remained friends for years to come, and much later, Fregin would share the future with Lazaridis.

Lazaridis excelled in every subject at school, but especially in the technical ones. Among these, electronics was particularly appealing. He and Fregin built things together, and sometimes were recognized for their creations. In fact, a children's author mentioned one of their pursuits in her book. "The two boys entered the high school science fair. Their project, a solar-powered water heater that could track the sun, won the city-wide Windsor Science Fair."[1]

One of Lazaridis's teachers in those years was a man named John Micsinszki. He was probably the first person to recognize the boy's attributes and potential. Years later, he recalled their interaction: "Mike first came to my attention when he was in grade nine. He was very bright and had a thirst for knowledge. He was exceptional. He had a very strong work ethic, an excellent sense of intuition, was a fast learner, and problem solver."[2] For those reasons, the teacher allowed Lazaridis to pursue several avenues of study, many of which were not on any curriculum, but Mike learned from them. He built things, repaired things, envisioned things from television to electrical circuits, to computers. In the latter lay his future.

While he was in high school, Lazaridis heard about the first computers, what their purpose was, and what they could do. He read everything he could find about these revolutionary machines and in due course figured out how they worked and why they worked. Then he turned to Doug Fregin and told his friend that they were going to make their own — and they did! Its functions were somewhat limited, but the thing could calculate, add, subtract, and solve simple problems. Both boys were naturally quite pleased with what they had accomplished. Along about that time, John Micsinszki gave Mike some advice he never forgot: "Don't get too hooked on computers," the teacher pointed out. "Someday the person who puts wireless and computers together is really going to make something."[3] The comment was prescient indeed.

By the time Mike Lazaridis finished high school, his marks, work ethic, and drive would have taken him far in any university in Canada — or elsewhere. However, because neither he nor his parents had the financial wherewithal to enable him to study without the thought of the cost, an innovative program at the University of Waterloo, in Waterloo, Ontario, came along at just the right time. The students there could attend lectures for four months, then work for four in related fields, and at the end of the requisite time period earn their degree in whatever field they were pursuing. The idea was called Co-op (co-operative education), and it was a success from the outset — even though there were plenty of traditionalists who faulted the plan.

The author experienced this first-hand. At the time, I was a high school vice principal in Ontario, working in an academic environment, surrounded by academics. Most had gone to and graduated from universities in Canada, Britain, and occasionally the United States. All knew what university had been for them; some still revelled in the experience, others remained jaded by it. Yet virtually everyone felt, and strongly felt that you went to university and begged, borrowed, or relied on your parents to give you the money you needed to stay there. You did not under any circumstances, study for a while, work for a while, and get a degree that way. For that reason, far too many regarded the Waterloo experience as a somehow inferior one. In this they were wrong.

Where possible, the Waterloo young people spent the time away from the classroom working at companies where their interests and skills coincided. Often these companies needed help, and bright, keen, university students were just that help. On the other side of the coin, the co-op participants learned first-hand how business worked, and in many cases received job offers from the firms that took them on. As well, there were occasional instances where the temporary worker learned that the place where he or she spent time at was *not* the field of endeavour they might have thought it was. Occasionally, co-op placements were not up to the Waterloo standard, and were dropped. Nevertheless, it was into this environment that Lazaridis plunged. With the limited financial support his parents could offer, he was able to live in residence during his first term at Waterloo. The experience was beneficial in a variety of ways.

Lazaridis found himself among like-minded individuals, all of whom were interested in the kinds of things he was. This, he realized, was not like high school, where those in the academic streams got all the glory, won the most prizes, were the most popular, and so on. In contrast, those in the shop classes, most of whom were male, too often suffered in contrast. Now, at Waterloo, in the studies he did, it was the technical types who really succeeded. "Here," he noted in a particularly observant summation, "the most popular students were the ones who knew how to run the labs. I knew how to run the labs, so it was a different world."[4]

In that first term, Lazaridis studied electrical engineering and computer science, and learned all he could about the new machines and what they could do. He worked hard, did well, and by the time he was in his second year at Waterloo, was among those who were undoubtedly destined to succeed. There too, he had more of a chance to follow-up on the studies that interested him most. He did so with other bright young people who were out to make their world a better one.

Lazaridis went out on a work program with a company in Mississauga, just west of Toronto. There he absorbed much computer technology and gained first-hand knowledge of networking and various operating systems. He was also exposed to office politics, profit margins, personal and professional jealousy, and the ramifications of poor company decisions. Some of these resulted in the loss of key professionals from within the firm, and that bothered Lazaridis greatly. He resolved to learn from it, and much later explained his take on the subject. "There are a small number of visionaries within an organization and they're important, they're key developers; you don't want to lose them."[5]

His university studies continued, and Lazaridis matured. His familiarity with computers was an invaluable asset to him, and when he met and worked with others with the same interests, his world expanded. From an early vantage point, he dreamed of starting his own company. In it, he would have the control necessary to make it work, and he would be able to shape its direction. However, he really did not know then what such a company would do, what it would or would not make, or who might be involved. He just knew within himself, that he had to get on with it. He was not a sedentary visionary.

After a time in residence at Waterloo, he moved out, and, like so many university and college students in hundreds of places, lived in a variety of off-campus dives that were more functional than aesthetic. And while such places were party places for many, they were not so for Lazaridis. He was not much of a drinker, nor for that matter were most of the friends he hung out with. They were all more wrapped up in ideas — particularly the next big idea, wherever it might originate.

There were echoes of this sentiment many years later when, as one of Canada's most successful inventors, Lazaridis built a massive retreat on Lake Huron for himself and for the thinkers of the day who would come to visit. In showing the partially-built structure to the press in early 2012, he talked of his hopes for the beautiful place. What he said then could well have been his philosophy as a young student who looked into the future and wondered what it would bring. Of the visitors to his retreat, his purpose he said would be "to inspire the groups to think and think big and be bold." Then, with reference to himself and his achievements added: "That's something that has worked for me for decades, and I can't see it stopping."[6] And that was his mindset as he discussed and dreamed with his university colleagues when he was in his early twenties, and was living in Waterloo.

He still got together with his old friend from grade six, Doug Fregin, and together they built a few rudimentary things, including a device they called Budgie, which would display letters on a television monitor for commercial in-store placement and marketing purposes. But Lazaridis wanted more, and quickly convinced Fregin, who by this time was at the University of Windsor, to work with him. In short order, Fregin left his studies and came to Waterloo. At about this time, too, Lazaridis dropped out of university so that he could devote all his time to whatever it was the two of them could build. Then they founded a firm called Research in Motion, or RIM. It was incorporated on March 7, 1984, and in time would succeed beyond their wildest dreams, but falter, as well.

The first few months of RIM's existence were frugal ones. The founders won contracts to build advertising signs that were light-emitting diode, or LED panels for a General Motors truck plant. As an adjunct to this, Lazaridis bought a broken computer from the university, repaired the thing, and used it for much of the work the fledgling company was doing. "That computer never failed," he said later. "I knew how stable it was because I'd used it at the university. I was able to take apart the operating system and add my own pieces. I built a network for it; this was in 1984–1985 when nobody knew what networks were. That network became the forerunner for the signs and the technology that came later."[7]

Lazaridis and Fregin continued to struggle, but they threw themselves into their work and gradually saw results. After a time, they even began giving themselves modest salaries, and Lazaridis acquired a small car. It was a leased Honda, for which he paid about $150 a month, but it proved to be adequate for his needs at the time. During this period, he was busy writing computer code for what was called the CDS-100 sign system, as well as keeping abreast of changes in the business world, and in particular within the companies where he had contacts. RIM was by then receiving royalty cheques for each electronic sign that was sold. A big break was first one, and then a second, to General Motors.

In the months that followed, RIM did business with the National Film Board of Canada and was aided in this work by the advent of a microprocessor made by Intel. The NFB project involved building an electronic counter for their film synchronizers. The finished product was called the DigiSync Film KeyCode, and it was so successful that it was a winner in 1998 in the technical awards section for all that was best in the motion picture business. Lazaridis received an Oscar, and sales of the new product brought much needed revenue to RIM.

With the advent of wireless data systems, network paging, mobile messaging, and so on, an entirely new electronic age came into being. More and more people began to hear of something called email, even though initially few knew what it was. This was not the case with Lazaridis. He understood this world as did few others, and he moved forward with it. RIM became more and more successful, but it was rapidly growing even beyond Mike Lazaridis's personal abilities. "He had hardware engineers and software developers and innovative thinkers who could solve any technical problem, but he badly needed someone to help him with the business side of operations."[8]

To do so, he hired a man named Jim Balsillie, a Canadian who, in addition to being a chartered accountant, had a Bachelor of Commerce degree from the University of Toronto, and an MBA from Harvard. Balsillie was a hard-driving keen guy who threw himself into everything that took his fancy, and succeeded most of the time. The same age as Mike Lazaridis, he was equally bright, and proved it in his studies, in an accounting firm he worked for in Toronto, a financial one in Boston, and then with a Kitchener, Ontario, company. He left the latter to join RIM.

The first wireless products were less than successful, despite a succession of publicity spurts in various places and in several companies. But despite the lacklustre results, Lazaridis decided that RIM would phase out other operations it already had, and focus exclusively on wireless opportunities. By this time, he had named Jim Balsillie co-CEO of the company, and was adding more and more visionaries who would help carry his firm into the future. His vision was "to invent and manufacture user-friendly technology that people could easily carry and use to send and receive email."[9]

The results did not come right away, but they did come. In the time frame in question, there were problems, large and small; from the size of the case for the unit, to the type of batteries that would be inside it. The goal was to come up with a wireless, hand-held device that would work no matter where or when it was used. In the end, as we know now, Mike Lazaridis did just that. He had help, lots of help, financing, and luck along the way, but the best-known product he ever produced came into being. It was called BlackBerry, and it was launched in the third week of 1999. The product was a hit from the outset.

Mike Lazaridis at the launch of BlackBerry Bold in London, England, in July 2008. At the time, his company was riding high in markets around the world. Since then, BlackBerry has had its troubles.

Soon, thousands of devices were sold, then millions and millions, in Canada, the United States, and in countries around the globe. They were used by everyone from the guy who delivers your pizza to the man who would become the president of the United States. It is said that Barack Obama was never without his.

The revenue stream made Mike Lazaridis, Jim Balsillie, and so many others rich beyond words. The co-CEOs then ploughed vast sums back into the community and the country they called home. Their success continued unabated for an extended period of time, but then RIM began to experience a slide in market share that at this writing is still being felt. New personnel are at the helm of the company today. Balsillie has gone, but Mike Lazaridis is still on the board, though only one of many now. The time may be difficult, but this man, this Officer of the Order of Canada is not done yet. He has weathered tough times in the past, and many feel that he will survive this downturn, too. Lots of investors certainly hope so.

4

HENRY WOODWARD AND MATHEW EVANS
Two Friends Lighting the World

When Henry Woodward was only a child, his father Marc died of cancer. The family was distraught, but they had to carry on. Fortunately, because she was a neurologist, Henry's mother had the necessary wherewithal to raise her four children, and from all accounts was quite successful in doing so. Presumably, she must have been pleased later on when Henry, the oldest, decided to follow in her footsteps. As soon as he finished secondary school, he enrolled at the University of Toronto to study medicine. He was there when he chanced on the discovery for which he should be remembered, but sadly, is not. Most people today have never heard of him.

Woodward and his hotel-owner friend, English-born Mathew Evans, invented a product that should surely have brought them fame, fortune, and historical recognition, but that was not to be. Their work, the dedication, the time, the genius — and the luck — that it involved could have made them known, not just within their own country, but to the world at large. But that did not happen, for a reason that has been all too familiar for Canadian inventors.

They lacked the necessary finances to exploit their idea.

The two young men spent a good deal of time together. They lived in close proximity in downtown Toronto, and even though their paths in life were divergent, they had much in common. Most of all, they loved making things, doing experiments together, then talking at length about whatever caught their attention at the moment. Sometimes their friends laughed at their antics and their ideas, and even regarded them as "cranks" because some of their creations were thought to be rather ridiculous. In fact, when they came up with their great invention, those closest to them scoffed.

The discovery had come about by accident, and by the time their friends were laughing loudest, Woodward and Evans had their idea pretty well worked out. In reality, the matter had its origins some months earlier.

Most summer evenings, they loved to spend their spare time doing experiments with industrial batteries, induction coils, pieces of tin, clamps, and an assortment of wires. They always had pliers, screwdrivers, snips, and other tools at hand. Often the two friends would hook wires to batteries and watch to see what, if anything, would happen when they touched the wires together, or to various things at hand. On one occasion, just as it was getting dark, they noticed that every time they made a particular connection at the contact post of an induction coil, they could produce a rather substantial spark. They tried the move again, and again, and each time there was a spark. They played with the thing for a while, and congratulated themselves on what they had done. It had been fun, but initially, they did not give it serious thought.

However, the more they did the move, the more reliable it seemed to be. They then realized they could produce the spark *every* time. That realization caused them to wonder if making the spark — and producing some light with it, might even be practical. And it was not just about the spark as such; it was the light it produced that intrigued them. In an era of gas lamps, lanterns, and candles, this production of light was significant.

We have no idea today which of the men voiced the thought, but one of them wondered aloud that if the spark could be contained, or enclosed, would it be of any practical use — anywhere, for anything — perhaps even lighting a small room.

In order to test the idea, they needed an adequate location. At the time, a company called the Morrison Brass Foundry was in operation at 76 and 77 Adelaide Street in downtown Toronto. This establishment had been at that address for years, and later on had a branch plant in the west end of the city, in what is now Etobicoke. The Adelaide factory is long

Photographer unknown.

This is a rare photograph of Henry Woodward, Toronto-born co-inventor of the light bulb. No picture of Mathew Evans could be located.

gone now, of course, and has been supplanted by one of the major towers that line the street in that central part of the town. But, back in 1873 or thereabouts, the Morrison foundry was a going concern. It would prove to be ideal for what the two curious friends would require. There was plenty of space and all the necessary equipment.

Over time, they lugged out a large battery, wires, a carbon rod filament, and a water gauge from a boiler. They acquired two electrodes, and fastened one of these to each end of the carbon strip. This contrivance was then placed inside the glass gauge, and subsequently, the air was extracted from it, replaced with nitrogen, and sealed. Once the apparatus was hooked up to a battery, the carbon rod in the tube could be made to glow. This, in effect, was the crude form of the Woodward and Evans electric light. They refined the creation many times, and just as often, were able to make the carbon strip emit light.

And the naysayers continued to scoff. They came, out of curiosity and nosiness, to see what Woodward and Evans were doing, but instead of being supportive, left laughing. "Who needs a glowing piece of metal!" some said.[1] Others were even more disparaging, but the inventors were not discouraged. Instead, they began to seriously consider what to do with what they had discovered. After much discussion and soul searching, they decided to see if they could obtain a patent for their device.

That conclusion was a good one.

On July 24, 1874, they applied to the Patent Branch of what was then the Canadian Department of Agriculture in Ottawa, described in longhand the specifics of their invention, and attached hand-drawn illustrations of the device. Then they waited for a time, but in due course, received the results of their initiative. The two were granted a patent for their light bulb. The Canadian patent number was 3738, and the inventors rejoiced when they got it. In their application, they wrote that they had "jointly invented new and useful improvements in the art or process of obtaining Artificial light by means of Electricity."

But now the hard part began.

They had to exploit their discovery. In order to do so, the two approached friends whom they felt might be potential investors in the development of the light, and its subsequent marketing. After a few months, they had amassed a small amount of money, but once acquaintances had been approached, selling the idea to strangers was ever more difficult. However, inside of several months they had built up a modest financial base. It was getting to the next stage that proved to be more problematic.

At this point, Woodward alone felt that if the idea of the light bulb could be taken to the United States, a much larger and potentially more risk-taking group of financiers could be sought out. Because the two friends had been "unable to interest investors in their idea" in Canada, Woodward went on alone and "applied for a patent in the United States."[2]

Again, there was a waiting period, but finally, Woodward obtained what he wanted. He was given American patent Number 181613, for "Electric Light," and it was granted on August 29, 1876. Two witnesses to the acquisition were George Smallwood and John Roby, both of whom signed the declaration proclaiming it, as did Woodward himself and a lawyer named John Halsted. No doubt Henry Woodward now believed he was well on his way to financial success and critical acclaim. However, this was not the case.

Three more years passed, but because the necessary revenue had still not been realized, the Woodward idea languished. That was when a saviour with connections entered the picture. The person in question was a man in his early thirties, who in time would be recognized as one of the pre-eminent inventors/entrepreneurs/tycoons in the United States. His name was Thomas Alva Edison, and he learned of Woodward and Evan's creation and was intrigued by it. He decided to investigate further because he saw both commercial and financial possibilities for the device. To that end, he contacted the inventors of the light bulb in the hope that together they could arrive at some understanding relative to its development. Edison became aware that the two were Canadians, so he may well have touted his own Canadian roots when they met.

Thomas Edison's father Samuel was born in Canada, in the village of Vienna, Ontario, but as an adult, became involved in what was known as the Mackenzie Rebellion of 1837, an insurrection against the government of Ontario. However, "when the uprising failed, Samuel had to flee for his life … to the United States."[3] He settled in a place called Milan, Ohio, and even though his son Thomas was born there, the boy soon moved on. By the time he was sixteen; young Edison was living in Port Huron, Michigan, and was hired by the Grand Trunk Railway company that operated in Michigan, southern Ontario, and elsewhere. He was with that firm when he was sent to Stratford, Ontario, as a telegrapher. By all accounts, he was regarded as a satisfactory employee until he almost caused a train wreck.

Today, there is a commemorative sign in the local train station that mentions Edison's time in Stratford. A brass plaque erected by the city on April 1, 1940, states that he lived and worked in the town from 1863 to 1864, but not surprisingly, does not mention the incident that caused him to lose his job there. The matter had a rather comic resolution, but it could well have ended in tragedy.

One night, Edison was the sole telegrapher on duty at the station. In the course of his shift, two trains, a passenger and freight, were approaching Stratford from opposite directions. As part of his job, he was to signal the freight and get it moved onto a siding so that the other, faster train could go on through without stopping. However, Edison was still trying to contact the engineer of the freight as it went rumbling past his position. Now, two trains were steaming towards each other, both on the same track. However, luckily for all concerned, the landscape around Stratford is flat as prairie, and the two train engineers saw each other in time to stop. No accident occurred, but the matter led to Edison being ordered to appear at the railway head office in Toronto for disciplinary reasons.

Once there, Grand Trunk General Superintendent W.I. Spicer exploded with fury at Edison over what he had done. Much later, the telegrapher described the dressing-down he got that day, and that he was told he could well have been sent to prison in Kingston on account of his negligence. However, right in the middle of the tirade being directed his way, a chance occurrence presented itself, and Edison reacted immediately:

> Three English swells came into the office. There was a great shaking of hands and joy all around. Feeling that this was a good time to be neglected, I silently made for the door; down the stairs to the lower freight station, got into the caboose going on the next freight, the conductor who I knew, and kept secluded until I landed a boy free of fear in the U.S. of America.[4]

Once back in Michigan, the young man had no desire to return to the Grand Trunk in Canada, although he remained with the company for a while longer in the United States. By this time however, he was ready to develop his many other skills. He was creative, inventive, and always persistent, and he loved nothing better than to spend many hours alone in his workshop, building things, trying out new projects, and then securing patents. In his lifetime, he is said to have acquired well over one thousand.

One of his successful ideas was the invention of what was called the phonograph, or record player in 1877, a machine that enabled sound to be captured on a rotating disc and then played for others to hear. He made a great deal of money from it, as he did from many other ingenious creations. It was because of his love for new and useful devices that he had first heard of Henry Woodward and Mathew Evans and the light bulb they had invented. Ultimately, about five years after the two had come up with their idea; Edison decided that he would attempt to exploit it. He got in touch with the two Canadians, and following a series of rather brief negotiations, bought the rights to their invention. Today, Edison is often cited erroneously as the inventor of the light bulb, when in reality he merely purchased the technology. Unfortunately, he never really tried to dispel the idea that someone other than he had been responsible for this revolutionary device.

For example, when reporters and others asked him about his inventions, he was always rather coy when the subject of the light bulb came up, and he often split hairs in the way he explained its origin. "If I didn't develop the incandescent light bulb, then I didn't develop anything,"[5] he said often. This rationale was repeated, and led directly to Edison alone being credited as the inventor. That conclusion is still with us. Its entry on Edison in a popular American encyclopedia states without equivocation: "He invented the incandescent light bulb."[6] And that implication was certainly intended in a NBC television news story on September 23, 2010, about a light-bulb factory that was about to close for good in the United States. News Anchor Brian Williams highlighted the piece with the words: "It started with Thomas Alva Edison … and it ends tonight."

And the idea of Edison the inventor has often been repeated, without qualification, in Canadian publications, as well. In a travel article about the great man's winter home at Fort Myers, Florida, there is much about the subject, including the phrase: "one of his inventions — the electric light bulb."[7] However, more recently, the fact that Edison may not have been the inventor is reflected in at least two publications, both American. One of these was in a biography of Edison, when the author writes: "the inventor of *practical* (emphasis added) electric light."[8] Yet another, in a review of a book about Edison, there is a reference to him, but with the codicil: "The incandescent lamp he *perfected*."[9] (Emphasis added.)

So whatever the credit, the fact that Woodward and Evans were given some acknowledgement for their efforts must be mentioned. The amounts, as reflected in various publications range from $5,000 to $50,000, but the former would seem to be more accurate for the time. At any rate, Edison did pay to use the patents that had been registered by the Canadians.

In later years, the two inventors parted company. A disappointed Woodward left Canada for England, and apparently took up residence there. There is no further record of him. It is known that Mathew Evans died in Toronto in 1899. The two were on to something great, but neither gained much from it.

5

JAMES NAISMITH
A Round Ball and a World Game

James Naismith, who invented one of the most popular games in the world, had a tragic childhood. He was barely nine when his parents died, but he overcame the loss and learned from it. Within a few months, his beloved grandmother who had taken him in died too, but the passing of his younger brother at eighteen hurt the inventor more than he likely ever admitted. Yes, Naismith had so much to overcome in his first few years that his lifetime of achievement was quite extraordinary.

Born near Almonte, Ontario, on November 6, 1861, to Margaret and John Naismith, young James had a pleasant early life. His father owned a sawmill, and, as they lived in what was largely a rural area, James was able to explore and to run free with his friends in the fields and woods near the family home. But because he was the oldest in the family, he often looked after his brother Robbie and sister Annie when their mother was busy with the household chores that had to be done every day. Life in pioneer Canada never afforded much relaxation; all family members, no matter what their age, were expected to pitch in and help where they could.

But James's happy, carefree time came to an abrupt end. First, his father's mill burned to the ground; then a typhoid fever epidemic swept the area, and the Naismiths grieved as friends they had known for years contracted the disease and quickly died. Suddenly, and with almost no warning, John Naismith also became ill. Margaret did all she could to nurse her husband back to health, but instead of getting better, he got worse and died. Sadly, Mrs. Naismith soon took sick as well. She passed away "three weeks after her husband on November 6, 1870, James Naismith's ninth birthday."[1] The three children were taken to live with their grandmother until her death; but for the next years, were raised on the farm of their bachelor uncle Peter Young.

This time was no easier than the recent past. Even though the children attended school, there was always farm work to do in the morning, and after they got home every afternoon. And even though she was still very young, Annie did the cooking, while the boys helped their uncle in the barn and in the fields. In later life, James sometimes recalled milking cows, cleaning stables, and forking hay. He was doing a man's work when he was barely thirteen, so it is perhaps not surprising that succeeding in school became difficult for him. He began to struggle, fell behind, and dropped out in grade ten, at age fifteen. For the next several years, he worked on the farm in the summer. In winter however, he got a job elsewhere, but the labour that was involved was no easier. He became a logger — and that life was rough.

James Naismith grew up quickly as he lived among men, plodded through deep snow to cut trees, drove horses as the logs were hauled from the bush, and wielded axes and crosscut saws with the strongest men in the lumber camp. He grew taller, more muscular, and learned to drink and swear as readily as those with whom he worked. But this kind of carrying on came to a head one night in a tavern where a stranger recognized him and told him that if his mother was alive, she would be ashamed at the way he was acting. The comment startled Naismith, and he felt it deeply. But, in a testament to his determination, he took the remark to heart, put the whiskey behind him, and never drank alcohol again. At almost the same time, "he made another decision. After five years of drifting, he returned to school."[2]

He approached Peter McGregor, the stern, intimidating high school principal who remembered Naismith from talks they had had before he dropped out five years earlier. Now, the person who stood before Mr. McGregor was no longer the fifteen-year-old dropout; he was a twenty-year-old man. Fortunately the principal saw determination in his visitor's eyes, and after some discussion, agreed to admit James to his school. The decision was a wise one, and Peter McGregor never regretted making it. Nor did Naismith. In later years, he often reminisced about this period of his life, and was able to make use of his own experience whenever he had to deal with the recalcitrant or inattentive young men he would encounter in his later career.

But returning to the classroom was not easy. James found himself learning with young people, while his own youth was well behind him. Because he was an adult now, he was bigger than anyone else in the room, and sat at the back of every class he took. However, now he was motivated in ways he had never been earlier on. He listened attentively to the teachers, worked in a way he had never worked before, took materials home, and studied for hours by the coal oil lamp at the kitchen table. When he did not have his head in his books, he was with his uncle, helping with the milking, or feeding and watering cattle. On winter mornings, he was up early, working in the barn for a couple of hours before dawn. Then he quickly washed, had breakfast, and headed for school. His days were long indeed.

However, those difficult years paid off, and their conclusion came surprisingly quickly. "Jim was twenty-two when he obtained his high school diploma." He needed it "because he wanted to go to university."[3] And, according to some who have closely studied his early life, going away to school became somewhat of a hurdle. Peter Young needed his nephew on the farm, as fathers have needed sons on family farms since settlement in Canada began. Historically, when individuals such as James Naismith leave the country for a city — any city — they rarely return, and Young knew this. It was only after James promised to take courses leading to the ministry that his uncle acquiesced in his departure. Previous generations of the Naismith family had been staunch Presbyterians, not only in Almonte, but going back to their roots in Scotland. Peter Young was no different, and despite his nephew's brief moral stumbles in the logging camps, James "had been brought up to be very religious,"[4] and as an adult would continue to be a man of faith.

Finally, in the fall of 1883, the young man from Almonte boarded a train for Montreal, and for the first time arrived on the leafy and picturesque campus of McGill University. Then, apart from relatively brief visits to the Young farm, he would continue to live in urban settings for the rest of his life.

Those first weeks at McGill were not easy for the young man from the country, as he knew little about city life or university expectations. While he was not shy, he was reserved and often felt out of his depth. And as the campus at McGill was practically in downtown Montreal, the temptations of the town were readily at hand. However, Naismith resolved to stick to his studies, and did so. He rarely missed a class, and when he was not in a lecture hall, he was working in his room or in the university library. It was only after some weeks that fellow students whom he got to know drew him out of his self-imposed isolation. They introduced him to some of the sports at McGill, and he readily adapted to them. He became a gymnast, and a good one, and he played lacrosse and loved it. In fact, during his second year at university, he was a valued member of the Montreal Shamrocks, a world champion lacrosse team. But another sport appealed most of all.

James Naismith discovered football, and it became part of his life for the duration of his time in Montreal. He took to the game even more enthusiastically than he did to lacrosse, and he excelled on the field. And while he never had the huge stature that we associate with football players today, he was nevertheless just under six feet tall and weighed about 170 pounds. However, he was a natural and gifted athlete, and he quickly gained the respect of everyone with whom he played. During this time, he is said to have invented the football helmet.

This discovery came about for a perfectly understandable reason. Naismith grew tired of having his ears scraped and torn during some of the scrums that are integral to football. A day or so after one particularly rough encounter, he devised and wore what has been often cited as the first helmet ever worn during a football game. He is said to have told his teammates that he wanted to protect his ears, and while it cannot be cited definitively, the story is generally included in most accounts of his time at McGill.

But whether he really did come up with the headgear or not is probably immaterial. He quickly achieved a starring role in the game, and the skill he exhibited there carried over to other sporting venues. That in turn caused him to be noticed by university officials, with the result that Naismith was selected to become the first official director of athletics at McGill. The fact that this came about when he was still in his early twenties is a testament to his acceptance and initiative. He was involved with football, soccer, lacrosse, track, and other sports, but when away from athletics — studied and succeeded academically. However, such success and the happiness that resulted from it contrasted sharply with the devastation of a personal loss that affected him deeply.

His brother Robbie became seriously ill.

James was back for a short Christmas visit at Almonte at the time, and during those few days, Robbie began to complain of a stomachache that would not go away. There were no medical facilities at hand, nor was there any doctor who could be summoned to the home. In a matter of hours, Robbie's pains became worse, to the point that he could not leave his bed. James, Annie, and Peter tried to attend to the terribly ill young man, but their efforts were of no avail. Robbie's suffering quickly became more acute, and he passed away before their eyes. The cause of death was a ruptured appendix.

A few days later James Naismith went back to Montreal, utterly devastated. His parents, his grandmother, and now his beloved brother were all taken suddenly and cruelly. Now, it was his deep faith that gave him grounding. His university friends and associates did what they could, but only time became a healer. He threw himself into his studies and athletics, and continued to do well at university. Then, as soon as he had his undergraduate degree from McGill, he took the promise that he had made to his uncle to heart, and began studying to become a Presbyterian minister. He obtained his certification as a cleric in 1890, but almost immediately puzzled those who knew him when he told them he did not think he was meant to be a man of the cloth. Instead, he decided to mesh his two leanings: religion and athletics. He would continue to be involved in sports, but would incorporate his religious training to assist athletes in their personal life struggles. And he was true to his word for the rest of his illustrious career.

The year he became a minister, a somewhat unexpected opportunity arose for him to pursue his personal philosophy. He became aware of a two-year course that was being taught at a Young Men's Christian Association, or YMCA, in New England, and he thought it might be something he should pursue. After some correspondence back and forth, he moved to the United States and enrolled at the YMCA International Training

School in Springfield, Massachusetts. He passed the course after two years and stayed on at the YMCA as an instructor. His athletic prowess had been noted, and he would put it to good use.

Those who attended the school were all young men who, like him, were drawn to the YMCA philosophy of associating with others in an overall Christian sense. Most were in their early twenties and were interested in various sports. There was football, of course, and track, soccer, and so on, and all of these were played outdoors. Naismith organized games, often played himself, refereed, drew up schedules, and did all he could to make the athletic experience at the school of value for those who attended. In general, he succeeded.

But there was a gaping hole in the program.

Aside from things such as gymnastics and wrestling, there were few sports that were meant for indoors. That left the long winter months without any major game comparable to the outdoor ones played in the summer. For that reason, there was a limited outlet for the pent-up energy of the football players. This fact was noted by Luther Gulick, the director of the school.

One day, he broached this subject with Naismith, and after some discussion directed the McGill graduate to come up with some kind of game that would have wide appeal and that could be played indoors. Essentially, it would have to be "a team sport that would keep his football team in fit condition over the winter months."[5] James Naismith took up the challenge, even though he initially had no idea what he would do with it. He soon realized how thorny the problem really was.

He realized that any kind of new game would have to be played in the small school gym, so there would be severe space restrictions. As well, whatever he came up with would have to be different from football or soccer, for example. Both were suited for outdoors. If they were to be attempted inside, there would be far too many players injured. Those games were rough; they needed lots of space, so nothing like them would do.

As the days went on, Naismith thought long and hard about what Gulick wanted, and rejected every idea that came to mind. Finally, in desperation, he realized that some kind of ball would likely have to be a part of any game plan. But what would be done with the ball? A goal of some kind would be needed. And then, what would the players do?

The gym was really too small for running with a ball, and tackling on the wood floor would be much too dangerous. Finally, he decided that perhaps the players could pass or throw the ball to each other, and then attempt to put it into a box at either end of the gymnasium. But what kind of box? And how big should it be?

Naismith took his idea to the superintendent of buildings, Pop Stebbins, who promptly informed him that he did not have any boxes. He did have two old peach baskets, however, and wondered if they would do. Naismith thought they would.

"I told him to bring them up," the inventor recalled, "and a few minutes later he appeared with two baskets tucked under his arm. They were round and somewhat larger at the top than at the bottom. I found a hammer and some nails and tacked the baskets to the lower rail of the balcony, one at either end of the gym."[6] The railing was quite high, and he knew his students were not tall enough to simply drop the ball into the basket. They would have to toss it up. But should the ball be small or large? Again, he thought about this, and ultimately decided that a large ball might be better, so he chose a soccer ball.

Then, in order to clarify the purpose of the game in his own mind, and how it might be played, he sat down and wrote out some rules. He listed the things that he felt would be important, added one or two, then crossed some out, and sometimes changed both the order and the wording. In the end, he came up with thirteen in all. He was not sure if there should be more or less, but finally decided that the ones he had would do for a start.

Shortly before it came time to try out the new sport, he took his list of rules to the school secretary, Miss Lyons, and she typed them out for him. Then, he posted the list on the bulletin board in the gym, and waited for the eighteen young men in his class to arrive. They came in; Naismith read the rules aloud, and the first game of basketball began! This was just before Christmas 1891.

Even though Naismith did not know what to expect at first, he was pleased that those who played that day "were interested and seemed to enjoy the game."[7] It gradually gained popularity on the Springfield campus and in no time schools across the state heard of it and wanted to try it themselves. The farther it spread; the more enthusiastic the reception.

This life-size depiction shows James Naismith as an older man. In his hands are a ball and peach basket, the two main ingredients for the game he invented in December, 1891.

Naismith was happy with the reaction, but at no time did he ever boast or brag about what he had invented. The new sport did not even have a name, but when someone suggested "Naismith Ball," he demurred. But one day, when one of his students suggested "Basketball,"

Naismith readily agreed, and the name stuck. Today, millions of people around the world play the sport.

There were changes, of course. The peach baskets were a good beginning, but they were always a mild nuisance. Every time a player scored a basket, somebody had to climb up on a stepladder and retrieve the ball. For a while, a small hole was made in the basket bottom, and a broomstick was pushed through the opening and the ball was poked out. That worked until a hoop was suggested. Later on, the netting that we are familiar with today was attached to the hoop.

And in time, there were more rules: many more rules. Yet the thirteen that Naismith drew up were never tossed away. They were retained by his family until recently. On December 10, 2010, at an auction in New York, the original rules were sold for charity. The man who bought them paid $4.3 million for the two pages[8] that James Naismith had jotted down the night before that first game. We cannot help but wonder what this modest inventor would have thought of that. Probably, he would have been completely shocked.

6

WILFRED BIGELOW
Heart Surgeon and Lifesaver

The winter of 1941 was a particularly cold one across the northern part of Canada. There were several severe storms as well, and a single one of these claimed seventy-six lives. It was within the context of this background that a young male patient arrived at Toronto General Hospital one day, suffering from frostbitten fingers. The man was a Native Canadian, and his home was far from the city, but he had initially been seen by physicians there. The doctor who saw him was in his twenties, and he later admitted that he was unfamiliar with the treatment of frostbite and its more severe complications. And this case was surely that. Gangrene had set in, and the patient's fingers would have to be amputated. And the effects of gangrene, "the death of tissue in part of the body,"[1] cannot be reversed. The diseased body part has to be severed.

The doctor who treated the stricken man was Wilfred Bigelow, then in his final months of medical/surgical training in Toronto. He would soon be in the Canadian Army, helping to save lives on the battlefields of Europe. When he first saw the patient's fingers, he felt totally unprepared for what he knew he had to do. "It was a distressing experience to amputate the fingers of a young man,"[2] he wrote later. However, he went ahead and did so of course, but then the patient surprised him with an unexpected request: the man wanted his fingers back.

Subsequent to the operation, the fingers were with the Pathology Department at the hospital, from which they ultimately would be disposed. However, Doctor Bigelow was so moved by the man's request that he went and located the fingers, and just before the patient left Toronto General, returned them to him. The man's beliefs were at the root of the unusual wish. He wished to take them home and bury them there. But the fingers could not just be tossed in a hole in the ground and covered up. They had to be interred carefully, with the tips pointing down, in soft soil. Then they would be gently covered. If these factors were properly carried out, the patient was certain that later on, he would not have to endure what has been called "phantom pain."

And such a condition is not to be dismissed. According to the experts today at the renowned Mayo Clinic, the discomfort is "pain that feels like it's coming from a body part that is no longer there. Doctors once believed that this post-amputation phenomenon was a psychological problem, but [now recognize] that these real sensations originate in the spinal cord and the brain." Doctor Bigelow was told later that his patient apparently never endured this discomfort because of the way he had buried his fingers.

The young physician in Toronto never forgot this particular operation, or the original cause of the man's problem. He admitted from the outset to having only a superficial knowledge of frostbite and the effects of cold on the human body. It was that lack of understanding that in time would lead him to research the matter. It was also how he came to be associated with the cooling of the body prior to and during surgical procedures. However in the meantime, Wilfred Bigelow went away to war, and coped as best he could with the horrific medical emergencies he faced there.

When peace returned to the world, he trained for a time in vascular and cardiac surgery at Johns Hopkins University Hospital in Maryland, and then returned to Toronto General as a heart surgeon there. Around this time, he began to seriously research cold and its effects on the body, particularly during heart surgery. An early area of inquiry involved the wild animal *Marmota monax*, but more commonly known as the groundhog. Bigelow spent much time studying this little creature, because "in nature hibernating animals retain great tolerance to low body temperature, and it seemed logical to attempt to wrest their secret."[3] In other words, he wanted to know how the common groundhog stayed alive during hibernation, when the ground temperature around it fell markedly.

Groundhogs dig burrows deep into the soil, and sensing the approach of winter, simply curl up and nod off when the cold comes. They can sleep contentedly for many weeks, without food, water, or warmth. In fact, even when the earth around them seems to be frozen solid, these small furry creatures survive. Then they emerge in the spring, start to eat again, and go about their summer routine.

After much difficulty, Doctor Bigelow obtained groundhogs for study purposes. Doing so was a trial-and-error enterprise as the animals, while plentiful, were not always co-operative or easy to come by, and then after capture, had a sense of resilience that was acute. They could chew their way out of cages, hide amazingly well, and exercise an antipathy towards humans that always required caution of the part of their handlers. A small group of dedicated individuals did the trapping for Bigelow and his team, but the initial process was too slow for research needs. For that reason, a so-called "groundhog farm," with up to three hundred animals was established in the open country near the town of Collingwood, Ontario. Enough groundhogs were found, and over the course of months of study, their hibernating habits contributed to medical science.

It was found that the groundhog could undergo extensive cooling in the lab, and then its heart could be opened and examined for many minutes. The animal did not die during the operation; in fact, it recovered quickly after being warmed up. It would then be kept under observation in the days that followed, and in spite of the open-heart surgery, would be as lively and as mischievous as ever. After considering the groundhog recovery, Doctor Bigelow wanted to continue his research in order to see, as he hoped, if similar cooling could be of benefit to humans prior to an operation. He later admitted that he and those with whom he worked never did specifically isolate the gland or gland secretion within the groundhog that actually contributed to its withstanding cold. It was the search for this essence however, that prompted him toward further study with humans. He progressed to this plateau by working for a time with dogs.

The team used anaesthetized dogs, and all of these animals were extensively cooled. The various procedures went well until one day a dog died on the operating table. In attempting to save it, heart and chest massage were tried, but to no avail. Doctor Bigelow recalled looking at the heart; it was healthy in size and colour, and to his eye seemed perfectly normal. He had no idea why it suddenly stopped beating. But then, a remarkable thing happened. Bigelow described the moment.

"Out of interest and in desperation, I gave the left ventricle a good poke with a probe I was holding. There was an immediate and sudden contraction that involved all chambers — then it returned to standstill. I did it again, with the same result."[4] In the next few seconds, the heart stopped several times, and each time it was poked, started again. Finally, it was beating on its own, and in doing so left Doctor Wilfred Bigelow with a wondrous realization.

He observed that a heart that was stopped could be started again if given some kind of jolt. Almost immediately, he began to wonder if the same kind of thing might restart a human heart. It so, he was sure the process would surely save lives. Within this realization was the kernel of understanding that ultimately led to the heart pacemaker! It may not have been a "eureka moment," but it was awfully close to one. From that instant on, the search for a pacemaker for the human heart began.

The quest was not easy, but in the end it was successful.

Doctor Bigelow and a colleague of his at Toronto General, Doctor John Callaghan, who had been present when the dog's heart stopped, both set out to learn all there was to know about pacemakers, and if in fact they existed at all. Both men were young, enthusiastic, and so convinced of the potential that could result from the accidental discovery that they resolved not to be dissuaded by whatever obstacles there might be ahead. Then they partitioned the path before them.

Doctor Callaghan spent hours and hours in library research, reading every scrap of medical and historical literature he could find, all to see if someone else had come up with a pacemaker of any kind, at any time. He did learn of some rudimentary experiments and rather ghastly machines that had existed; and a few of these had even been tried on humans. Most of the earliest had some electrical part to them, whether it was attaching electrodes to a patient's chest, having others take "a special electrical bath" while nude, or having to wear an electrical belt. None of these things did any good, of course, but they provided Callaghan with plenty of material to amuse his friends over drinks.

There was one particularly awful experiment that seems to have been carried out by a mad scientist, sometime in the mid-eighteenth century. The purpose of the idea was to bring a dead person back to life; and it was deliberately tried by the originator on himself — before he died. It involved putting an electrode into the anus; another in the mouth, connecting the two, and then attaching them to the posts on a battery. The inventor who used it on himself claimed he saw flashes of light when the wires were hooked up. There was no record of further use, so we can probably conclude that it did not raise a lot of the dead.

One vaguely worthwhile device came along in 1932, invented by a New York doctor named A.S. Hyman. It too had an electrical component, and was used mainly on guinea pigs. A needle that carried an electrical charge was inserted into a live pig's heart. It did not keep the animal alive, but it did make the heart beat more often. The only useful aspect of the thing was its name: Hyman called his invention a "pacemaker." Little was heard of it in the years that followed.

While Callaghan was searching the written records, Doctor Bigelow sought out someone who could make some kind of heart pacemaker for him. The doctor had a vague idea of what he wanted, but had no idea at all how such a thing might work, even if it was constructed. He took his search to Ottawa, and fortuitously linked up with a senior official at the National Research Council. Bigelow first had to state his case for him, and was quite convincing in doing so, even though the doctor was not at all sure if his request would find any kind of sympathetic ear. Fortunately, it did. He was told about an electrical engineer named Jack Hopps, who shortly thereafter came to Toronto and met with Bigelow and Callaghan at the Banting Institute at the University of Toronto.

The two medical men told Hopps the probe story, and how the dog's heart, even though it had stopped, was induced to start beating again after Bigelow poked it. They asked Hopps if he could build some kind of machine to do the equivalent of the poking, and assured him that if he invented such a thing and it worked, lives could be saved. Hopps not only agreed to see what he could do; he agreed enthusiastically. In short order, he began his work. He had a well-equipped laboratory in Ottawa, so he returned there to devote his full attention to the project.

Meanwhile in Toronto, there followed weeks of trial and error and operations on several dogs. All of this study period was important in order to ascertain what kind of electrical impulse would suffice, whether it would work as hoped, or if it needed to be abandoned, or reworked.

While Bigelow and Callaghan did their part, Hopps proved to be the electrical genius he had been touted by his superiors. He simply never gave up; even though there were times he surely must have been tempted to do so. Instead, he gathered a small group of others around him, told them the purpose of the inquiry, and set them to work doing what he felt Bigelow and Callaghan wanted. As many ideas were set aside as were attempted, but slowly, painstakingly, and determinately, they began to see the results of their labours. Through all this time, the two physicians at Toronto General were kept abreast of what was being done.

Bigelow and Callaghan worked with dogs for days on end. They sometimes stopped the hearts of anaesthetised animals, kept careful track of the time of the stoppage, then used massage or shocking to get them going again. At other times Hopps and Callaghan were together, along with several technicians, doing "a careful and painstaking series of experiments: to access the electrical activity on a normal heart; to determine comparable pulse characteristics that were most effective and safest; to decide on the best method of delivering a stimulus to the heart."[5]

Their efforts paid off, and gradually the heart surgeons had a pretty good idea of what was needed. And because they knew they had to come up with something without doing an open-heart operation on the dogs, they instead inserted a tiny wire through a cardiac catheter into a vein in the animal's neck, then through to the heart. Next, using electrical stimulation, they could nudge the heart itself. The procedure worked. Hopps took that data back to Ottawa and, in due course, used it to build the first pacemaker. The device was about the size of a boxed six-pack of beer.

There was still much more trial and error ahead and not all of the procedures that were followed were successful. However, the pacemaker, an outgrowth of Bigelow's accidental probe, had been invented. In the weeks and months afterwards, it was perfected and perfected again. The invention of the device and its use, coupled with the use of hypothermia were reported and much heralded in the scientific community. The announcement was made by Hopps, Callaghan, and Bigelow in a paper delivered at a gathering of the American Surgical Association at Colorado Springs, Colorado on April 20, 1950. Then that fall, on October 23 in Boston, the pacemaker and its potential was explained at the annual meeting of the American College of Surgeons. There were over nine thousand doctors there, from dozens of countries, and the news was greeted with much acclaim. For Bigelow, Callaghan, and Hopps, the acceptance was a resounding justification for all their work. They returned home from the conference with much satisfaction. As might be expected, the press picked up on the matter, and soon the three principals became known and lauded across Canada and elsewhere.

But Wilfred Bigelow and his colleagues refused to call that plateau the end of the saga. At the Colorado Springs conference for instance, they described the operations on four dogs: "In two experiments the artificial pacemaker was used for ten or fifteen minutes, and when it was discontinued the heart returned to standstill. In the other two animals following electrical control of the heart for ten and thirty minutes, normal spontaneous heartbeats returned."[6] But this testing had been on dogs. Treating human beings would be an entirely different matter.

In the decade after the first pacemaker, there were rather few times when it was used on humans, but the device was simply too large and too experimental. However, when transistor batteries became widely available, the problem of size disappeared. Tiny pacemakers could now be constructed and placed under the surface of a patent's skin. But at first, even though these devices quickly came into their own, there were lots of teething problems. Some of these included "faulty batteries, poor connections between the wire and the heart muscle, a leak of bodily fluids through the encasement, and broken wires."[7]

Doctor Bigelow called this time a "learning period," but it ultimately led to various solutions and then widespread use of the device. In time, thousands and thousands of heart patients around the globe owed their lives to this little electric machine. It had come about, as we have said, because a curious medical man had poked the arrested heart of a dog, and caused it to beat again. And as a curious footnote to the invention, John Hopps required a heart pacemaker himself in 1984, "[and it] was replaced thirteen years later when the battery weakened. He died in 1998."[8]

But just who was this Wilfred Bigelow? Where did he come from? What was his background? And more importantly perhaps, how was he regarded by those in his profession and others outside it?

The story starts on the Canadian prairie in the early part of the twentieth century. A man named Wilfred Abram Bigelow and his wife Grace Ann Gordon lived in Brandon, Manitoba, and raised their family there. The man whose idea led to the pacemaker was their son, born in 1913 and named after his father. And it was the father who was the lasting inspiration for the young doctor who wondered about the effect of cold, and performed that gangrenous amputation that he never forgot.

Wilfred Bigelow Sr. was a highly respected country doctor in Brandon. He was naturally skilled, inquisitive, hard-working and innovative in ways that were unheard of at the time. For instance, he often ordered the nursing staff who looked after his elderly patients to let them have some rye whiskey in the days following surgery. This, he felt would make them more contented — and presumably it did. The man always kept an open mind and never, ever, gave the impression that he had all the answers. That trait he passed along to young Wilfred who he did his best to live by such example. "Instead of impressing those about him with how much doctors know," Wilfred Jr. would write later, "my father was teaching an appreciation of the limits of scientific and medical knowledge."[9]

And every year, the father would leave Brandon and go to Winnipeg, Toronto, New York, and elsewhere to learn all he could about the latest medical advances and treatment techniques. He then would come home and incorporate the best elements in his own practice. He benefitted, his patients benefitted, and so did other Manitoba doctors who sought his advice. He also was the driving force behind Canada's first private medical clinic that functioned for several decades in Brandon.

And the son learned from the father. It almost seemed pre-ordained that Wilfred Jr. would go into medicine. He looked up to and loved his father deeply and that admiration and respect would later be reflected in the writings of the pacemaker pioneer. Young Wilfred got his elementary and secondary education in Brandon, and then went on to the University of Toronto for a bachelor of arts and his medical degree. Finally, after the military service and post-graduate work noted earlier, he elected to remain in Toronto for the balance of his career.

And what a career it was! Doctor Bigelow continued to be a pioneer in the use of hypothermia in the operating theatre and did all he could to assist young medical students in its benefits. A few years after the pacemaker initiative, he founded a university program specifically targeting heart surgeons. This was the first of its kind in the country. However, his "special contribution to surgery of the heart [was] the use of hypothermia to slow tissue metabolism and thus protect the heart and brain from damage."[10]

In later years, Doctor Bigelow was recognized for his lasting contribution to medicine. His inclusion in the Canadian Medical Hall of Fame was welcomed by those who knew him, those who worked under him, and those who had heard of him, no matter where they were. His official citation calls him "one of the most distinguished surgeons Canada has ever produced and stands among the world's titans of medicine."[11] On October 21, 1981, he was named an Officer of the Order of Canada as "an internationally acclaimed pioneer of hypothermia in heart surgery."[12] Doctor Wilfred Bigelow passed away on March 27, 2005.

7

PETER ROBERTSON
Small Hand Tool, Big Impact

One summer morning, a little over a century ago, twenty-year-old travelling salesman Peter Lymburner Robertson was building a wooden display stand at a function on St. Peter's Street in downtown Montreal. A handful of curious onlookers were on hand, even though they were not really interested in what he built; they were there to see this sales guy use something called a spring-loaded screwdriver. For a while, all went well, and the spectators continued to watch. But then, without warning, the little show came to an abrupt conclusion.

As Robertson put increased pressure on one of the flat-head screws he was inserting, the driver slipped out of the slot on the thing, and ripped into the palm of his hand. There is no record as to what he might have said just then, but as he wiped up the blood, the salesman decided that there had to be a better way.

And there was — or soon would be.

Robertson told the North Brothers of Philadelphia, his bosses at the time, that he was quitting his job and would be returning to his home in Milton, Ontario, and to his workshop there. This was because he believed that while he might have been a good salesman, he thought he would be equally good at inventing things. In short order, he proved just that.

After a number of false starts, failed experiments, and periods of utter frustration, the young man came up with a new kind of screwdriver, and a screw that was made for it. The fastener in question was unlike the common flat-headed one, with its matching driver. The tip of the driver that Robertson invented was square, and it could be inserted into a small, corresponding square on the top of each screw. Because the two fit snugly, there was no slippage, so hopefully there would be no more cut hands. As well, the recessed part of the screw head was tapered, as was the new kind of driver. Both modifications were important, particularly when the tool was used with one hand. A recognized fault of the old flat-headed screws was that the driver often slipped when they were being inserted. As well, anyone using the flat-head had to use two hands, at least to get the screw started: one hand had to hold the screw and the other to turn the driver. Then, if there was a slip, the old screw would often turn on its side, causing the point of the driver to damage whatever material was being fastened. If, for example, what might one day become a beautiful dining room table could well have its surface gouged if the driver slipped. There was little chance of this happening when Robertson's invention was used.

Now however, he had to show others that his idea was a good one, and the first step in that process was to obtain a patent for his brainchild. This took some time, but by the end of 1907 he had his first patent for the screw and its driver. This version was later modified, but it was the one questioned in a rather vindictive article in the August 1910 publication of a then-popular magazine called *Saturday Night*. Robertson responded to the criticism with a lengthy letter to the editor. He cited the modifications, won his point, and the complaints subsided. Soon, his socket-head screw gained even more acceptance.

The Robertson screw, as with all screws, is what the ancient Greeks and others called a "simple machine." Historically, there have been only five of these: the lever, the wedge, the wheel and axel, the pulley, and the screw. All other machines are really just complications of one or more of these.

The screw has been around since ancient times, and may well have been invented by the genius Archimedes, one of the most accomplished inventor/mathematicians in history. What the man from Milton produced was a modification of Archimedes' work: a new tool that was simple, lasting, and functional. No one before him had successfully done what Peter Robertson did.

And yes, there had been earlier versions of the Robertson idea, but these were both rudimentary and not particularly helpful. One such originated in about 1875, but it was never really developed. Others that came along were even less worthwhile. That was why Robertson sometimes referred to what he came up with as "the biggest little invention of the twentieth century." He may well have been right. However, exploiting his discovery was far from easy. As mentioned, some disputed his claims; others his patent; and others his chances of developing the manufacturing capability to make the screw and driver so cost-effective that potential users would be able to afford the tool in the first place. But none of these impediments deterred this determined Canadian. He always had been what might be called a self-starter. He was inquisitive, determined and hard working — and he had been like that since childhood.

Peter Lymburner Robertson was born into a farm family in Haldimand County, Ontario, the youngest in a family of six. It was said that he was much like his Scottish-born father, who by all accounts was a gifted entrepreneur. Sadly, the father died when Peter, or P.L. as he would later be called, was only seven. Robertson Senior was fifty-three when he passed away, and P.L.'s mother had to raise the brood alone. When P.L. was fifteen, an adopted son joined the family, and that meant one more mouth to feed. Nevertheless, Annie Robertson was a fine matriarch, and all seven children did well in life. Annie, who was also a Scot, was born not far from the Robertson farm, and in all likelihood may have met her future husband when she was quite young. In any case, she was an important influence on her youngest son, and in later life, he always referred to her with affection.

As a young widow, Annie Robertson did a remarkable job raising her children, particularly when there was little money around. Yet perhaps it was this Spartan existence that instilled in all the youngsters a strong work ethic, and a belief that if they wanted success in life, they had to strive for it on their own. All of them, P.L. included, had to struggle, but none lost the will to better themselves. They knew their mother was strict, God-fearing, and supportive. However she did not suffer fools gladly, nor did she tolerate the shirker. Her children all knew this, and were careful to avoid her censure.

Being a member of a farm family meant long hours tending live-stock, cutting grain, ploughing with horses, and cleaning barns in the winter. There often did not seem to be enough daylight hours. And there was disappointment, heartache, and loss. Sometimes crops failed, cattle got sick, or rain did not come when needed. In the winter, the roads were not ploughed, so the only ways to get around were by sleigh, cutter, or horseback. Often though, it made more sense just to walk to wherever you were going: to the nearest town for supplies, to the service at the local Methodist church, to the post office where mail arrived on days when there was no major blizzard in progress.

Yet P.L.'s formative years were generally happy ones. There were family gatherings, church functions, games of various sorts, and visits from friends and relatives. When he could, P.L. would spend hours by himself in a farm shed where he had a rudimentary workshop. He built toys and invented contraptions that sometimes never worked as he hoped, yet he did his best with the few tools at hand. It was during these growing-up years that he developed his love of all things mechanical. Whether other kids laughed at this solitary little inventor is not known, but in all likelihood, sibling teasing would not have deterred him at all. P.L. did things *his* way then, and later in life as well.

Shortly after he invented his marvellous new hand tool, P.L. set about establishing a manufacturing operation for it. Initially, because he did not think he would get any help near home, he began exploring sites further afield, from Hamilton to Niagara Falls and elsewhere. However, when the Milton Town Council offered him a $10,000 tax-free loan if he would build his plant in their town, Robertson accepted the terms. By this time, he had several investors interested, and together they went ahead and got the operation underway. Over time, the loan from Milton was paid off, and the P.L. Robertson Manufacturing Company Limited expanded and prospered. By this time, scores of contacts P.L. had made in his past sales jobs assisted in boosting the company's profile. There were growing pains, of course, but when carpenters and others saw how well the Robertson product worked, they were converted. More plant employees were taken on, and the new business on Bronte Street in Milton grew steadily. In fact, it is still there today — although there are now operations elsewhere as well.

The first years of enterprise were modest, but gradually P.L. expanded his company and made more sales. He worked tirelessly doing so, and insisted that anyone he hired be diligent and dedicated. And most were. Those perhaps who had less initiative were sometimes shamed into putting more effort into their work. P.L. spent long hours at the plant, and adopted a hands-on approach to everything he did. He was all over the facility, every day, and employees who might have been tempted to be less industrious often changed their work habits when they found P.L. standing beside them at times when they did not realize that he was nearby. P.L. Robertson was a good boss, but you put in a full day's work if you wanted to remain in his employ.

Gradually, the Robertson screw and driver gained acceptance across the country and turned up in manufacturing enterprises big and small. Certainly the largest of these was at the Ford motor company in Windsor, Ontario. There, the new and versatile Model T began rolling off the assembly line and Robertson screws held the car together. Several sections of the auto were made of wood, and plant employees found they could assemble the machine much faster if they used Robertsons instead of slot-head screws. They could drive the fasteners with one hand; there was almost no slippage, and they could finish the product quickly. In no time, cars on the Canadian assembly line were being built much more efficiently than those in Michigan, just across the Detroit River in the United States.

The increased productivity in Canada ultimately came to the attention of auto magnate Henry Ford himself. He ordered a detailed study of the contrasting assembly lines, and soon noticed the obvious. That was when he decided that Robertson screws and their drivers would henceforth be used in *all* his plants. But before this could happen, he had to approach P.L. and negotiate the matter.

Unfortunately, the ensuing discussions soon ground to a halt — and Ford had to accept his share of the blame.

P.L. was a proud man, proud of his fledgling operation, and proud of what he had accomplished. He was happy to sell to Ford, certainly, but only if the rights to his company remained where they were: in Milton, Ontario, Canada. Unfortunately, the headstrong Ford would have none of that; he wanted more. He wanted all the rights to P.L.'s company, and if he did not get them, he would not be using P.L.'s product.

P.L. was just as adamant that the rights were not for sale. Ultimately, he turned Ford down, but the Canadian company suffered in the long term as a result. One wonders in hindsight if P.L. ever second-guessed himself about the decision then, or later. Unfortunately, there is no definitive record remaining about his feelings on the matter. Incidentally, Ford later learned that a man named Henry F. Phillips of Portland, Oregon had come up with another design for a screw. This time, instead of a slot, or square recess on top of the thing, Phillips's invention had a small cross on the screw, and a matching one on the driver. From that time on, Phillips's cruciform gained wide acceptance in the United States. Today, that invention is much better known there than the Robertson ever was.

As might be expected in the aftermath of the Ford disagreement, the Robertson company suffered. However, P.L refused to admit any kind of defeat. Instead, he went to Britain, secured financial backing, and established a company called Recess Screws there. For a time, this fledgling firm had some success.

But not for very long. There were disagreements among the investors, discontent over future prospects, and disappointment for P.L. Ultimately, he decided to close the English operation and return home. Not surprisingly, the Milton firm was impacted, yet over the years, in spite of declining sales, worldwide recession, and plant layoffs, screws were made. Ultimately, the two major wars of the twentieth century proved to be saviours for the company. Robertson not only sold such things as brass rivets to the military here and in England, his company also made thousands of the high-explosive shells that were sought after in those times of turmoil. Then, at the end of the Second World War, Robertson screws were needed as the civilian economy expanded more quickly than at any time in history. In the 1950s, more than five hundred people worked at the Milton plant.

But what of Peter Lymburner Robertson himself?

Over the years, P.L. had shepherded his company through good times and bad. He rejoiced when things went well; he took it personally when he had to lay off employees; he gained stature in Milton and did what he could for his town and his country. Sadly however, he often failed to look after himself. He rarely relaxed, seldom took a holiday, and worried incessantly about his company, its shareholders, and its future.

P.L. never married, although as revealed in an excellent biography of the inventor, author Ken Lamb mentions that P.L. had a daughter. She was born to a housekeeper Robertson once employed, but the child's existence was largely hidden from the Milton populace. Later, as an adult, she always felt estranged from her father, and remained so until the day he died. The fact that his declining years were difficult likely did not help heal wounds.

P.L. drank, at home, at his favourite hotels, and on the job. Undoubtedly, he would be called an alcoholic today. For the most part though, he continued to run his firm, and run it well, even as his alcohol consumption increased. But he also had other health problems.

For several years before his death in 1951, at the age of seventy-two, he had been a diabetic. Unfortunately, despite remonstrance for doctors and others closest to him, he refused insulin treatment. His eyesight was going, and he finally had to be hospitalized under lock and key for the severe hallucinatory problems that he suffered. He was convinced that imaginary forces were out to get him. When his death came, it was undoubtedly a blessing for himself, for those who knew how much he had deteriorated, and for those friends who stood by him in good times and in bad. He had left his mark in life, and then was gone.

8

JOSEPH-ARMAND BOMBARDIER
The Man Who Conquered Winter

If you have ever ridden a train, flown in a plane, hopped on a subway, or zoomed along a winter trail on a snowmobile, you may well owe Joseph-Armand Bombardier a vote of thanks. His name is connected to all of the above. That is because he was the inventive genius behind the Ski-Doo snowmobile and it was the success of that revolutionary machine that in large part led to the multifaceted company that is called Bombardier today. The founder did not live to see what his business would become, but he laid the foundation for its future and inspired those who would take his life's work and build upon it. Today, thousands of people in several countries depend on one or more branches of the Bombardier conglomerate for their livelihood. So if ever there was a success story, this was it.

Yet the origin of the present-day empire lay in the rather comical adventure of a fifteen-year-old boy and his little brother. The year was 1922 when Joseph-Armand and Leopold Bombardier went for a wild ride on a homemade snow machine that had been built by the older of the two in a Quebec garage. The contraption had a rusty Model T Ford motor mounted on the rear of a full-size sleigh, "with a spinning wooden propeller sticking out the back."[1] Armand, as he was generally called, had fashioned and attached the propeller himself, and the engine used was an old one his father had given him. He had rebuilt it and, with his brother's help, attached it to the sleigh frame. Then the two boys opened the garage door, climbed on board their creation, started the engine, and roared away. They headed down one of the snow-covered streets in Valcourt, their hometown, and profoundly startled everyone they encountered. There was no muffler and that resulted in a sustained and loud racket that was heard all over town.

Within seconds, it also got the attention of their father — and he was furious. He ran from his house and raced down the street after his sons. By the time he got them stopped and the racket curtailed, he was of two minds over what had just happened. While he was profoundly upset at what they had done, as a father he was thankful they had not killed themselves riding on what they had built. There is no record today of how the machine was returned to the garage, but Alfred Bombardier flatly refused to let his sons ever ride on the thing again. So Armand's first experience with a snow machine was over. It would not be his last — although several years would pass before he would ride one again. By then he would be an adult.

As a child, young Bombardier was always intrigued by things that moved; things that worked, and mechanical things that he could build. "He made a toy tractor that moved on its own. He made a small cannon that fired. By the time he was thirteen, he knew how to fix broken engines" so well that "neighbours often called on him for help."[2]

In due course, his father gave him a small garage where he stripped down engines, rebuilt them, and put them back in cars, restored and almost good as new. He continued to build things, either by instinct, or from instruction books he accumulated. Often these were written in English, so he began studying that language, and over time became fluently bilingual in both French and English. For a time he worked and apprenticed in a Montreal garage, but then returned to Valcourt at age nineteen and opened a garage of his own. Because of his innate understanding of motors and how to repair them, his business was soon a success. Then, and for the rest of his life, he employed family members and others from his hometown. And, because work in the area was often scarce, those who were hired owed their livelihood to him.

The garage business, which included gas pumps out front, continued to thrive. After about three years, Bombardier married, and he and his wife Yvonne started a family. Well before marriage however, he used to think back to that first "snowmobile" he built and which he and Leopold had ridden that one time. And because it had been fun to build, and even more thrilling to ride, he gave more and more thought to constructing some kind of snow machine that would be practical — and safe — in the long winter months when the roads in his home province were often impassable. He was still fashioning one of his early prototypes when a tragedy occurred within his own family. He and Yvonne had a two-year-old son who suddenly took seriously ill.

A local doctor was called, and he quickly determined that the sick child had appendicitis, and needed an operation right away. Unfortunately, the doctor was not a surgeon, and had neither the equipment nor the training to perform it. And because the nearest hospital was in Sherbrooke, fifty kilometres away, the snow-packed winter roads made getting there impossible. The consequence of the matter meant that the sick child died. The Bombardiers were completely devastated, but the tragedy became an impetus for the future. Armand determined then and there that he had to find some way to travel quickly in winter, particularly in an emergency. If he had had any kind of snowmobile that would have worked at such a critical juncture, he felt that he might have been able to get to Sherbrooke and his son would have lived.

Today, because we see and are familiar with the snowmobile, it is easy to assume that these machines have been around forever. Such is not the case, of course, any more than it is for cars, planes, and all the other machines that we take for granted. Someone had to come up with the idea for them. Someone had to build the first one and someone had to modify and perfect it. Armand Bombardier invented the Ski-Doo snowmobile.

And his first experimental machines were the result of trial and error, elation and disappointment, dogged determination and lasting persistence. His neighbours in Valcourt often saw him testing strange vehicles that looked to be half-sleigh, and half-car or truck. By building upon the idea of that first conveyance that he and his brother rode, he continued to work toward what he had in mind.

And he did so in all kinds of weather, whether it was snowing or not. He simply needed snow and there was generally lots of it. Very early on, he decided that protection from the weather had to be considered for what he wanted. To that end, he built various types of housings for his creations. Then he would climb inside these things and test them himself — sometimes walking back to his garage afterwards, looking downcast, because whatever concoction he had been riding broke down, got stuck, or even toppled over. The process of invention was not easy, "but he was not easily discouraged. If a vehicle did not perform as he wished, he simply took it apart and kept only those features that he really liked."[3] Then he would rebuild the thing, often several times.

Those first snow machines were large, often unwieldy, and heavy, because car or truck motors were needed to run them. Armand put strong skis on the front of the machines, and then figured out the best way to steer them. Instead of wheels on the back, he used treads, smaller and lighter than ones he had seen on army tanks. And because he used treads, they did not sink into the snow the way wheels would have.

And as his business improved and his dreams of a snowmobile advanced, Armand became more and more methodical in the way he worked. "He would think of the new designs in the privacy of his office, make detailed sketches to show to his co-workers, and then they would offer suggestions before the team went ahead and custom-made all the parts."[4] And the parts they made included the wheels, the treads, the cab over the entire machine, and occasionally even the type of material that was used to make the cab.

Sometimes the machines worked well and could be operated with ease in the snow. At other times, if the snow was wet and deeper than usual, the weight of the snowmobile caused it to flounder. If that happened to him, Armand knew it would happen to future owners, so he tried in every way he could to maintain safety, but to decrease the weight. The same thinking went into the operation of the motor being used. If the motor worked perfectly for Armand, or one of his employees, but not for a prospective buyer, then that was unacceptable, too.

In all of the building, testing, modifying, and planning, Armand showed his true nature. He was a good man, but a tough boss. He was a perfectionist and he expected perfection in his employees. He was often impatient, generally cheerful, headstrong, and occasionally uncommunicative. But at other times, he could be kind, supportive, humorous, and self-deprecating. His workers knew his moods and adapted to them. They avoided him when he was in a rage over something, but asked his advice when he calmed down. Then, he was invariably helpful.

And Armand was always in a hurry. There are lots of stories of his racing up stairs, two or three at a time. Often he would start to show a worker a particular task, but if the employee was slow in catching on, Armand would push the person aside and do the job himself. Yet he could relax. He loved fishing and hunting, and, when time permitted, would go on trips to pursue these sports. He liked to be alone in the bush and often used to say that he did his best thinking there. In the woods, he was always at peace.

Gradually, his inventive efforts started to pay off. His company built a roomy, seven-passenger snowmobile that resembled an oversized Volkswagen bug. "It was driven by a rubber-cushioned drive wheel and track, and could travel over snow that stopped all other vehicles."[5] The machine was quite versatile and quickly became popular in Valcourt and soon elsewhere. In almost no time, news of this winter accessory spread across Quebec, to adjoining provinces, and into northern United States as well. And even though the inventor stood by his product and believed in it, he too was pleasantly amazed at the positive reception. He named the machine the B7, with the *B* meaning Bombardier, and the *7* referring to the number of passengers.

There are many older people in Quebec today who can still recall the launch of the first few B7s, in part because they became so well known, they lasted for years, and they served so many purposes. Doctors drove them when making winter house calls to remote areas. Police forces had them. So did bus operators, school boards, and ambulance providers. Priests needed them to reach distant mission churches. Lots of enterprising businesses put them to use: telephone companies, delivery outfits, logging companies. Soon, Bombardier was receiving requests for even larger machines.

By then, Armand no longer pumped gas. His garage still stood — and does today, as part of the J. Armand Bombardier Museum in Valcourt — but his work now was the manufacture of snowmobiles, and lots of them. He adopted a clever method of advancing sales and the idea was sound. He would often climb into one of his machines and head to places that were farther and farther afield. There are plenty of surviving anecdotes of his venturing into outlying towns, driving down the main street, and then parking in front of the local newspaper office. Naturally, the editor and any reporters who might have been there would come out to see, first-hand, this vehicle they had heard about, but may never have seen.

In March 1943, Armand Bombardier posed inside one of his early snowmobiles. This one carried several people. The small, personal snowmobiles did not become popular until the 1950s.

Then, after he watched the news people admire what he was driving, Armand loved to invite them to climb inside. They would do so, of course, and then he would take them for a rollicking ride out of town, up hills and down, through bush land, and sometimes even past their own homes. By the time he brought them back to the paper, he was pretty much guaranteed that there would be a story in the next issue, about this wonderful winter invention, what fun it was to ride, and how helpful it would be to *so* many companies. No wonder the B7 sold, and sold quickly and consistently. It also led to the acceptance of the next generation of machines.

The B12 came next, and it could carry even more passengers and go anywhere the B7 could. In due course, adaptations of some of these giant machines became staples in the Canadian North. Dubbed the Muskeg Tractor, it proved especially suited for travel in areas that were inaccessible to other kinds of movement. Companies that did oil exploration, road-construction outfits, and builders of ski resorts all loved it. In no time, "the Muskeg Tractor was being used in Japan, Africa, Alaska and Australia. It was even used to move sand in the Sahara Desert"[6] and as a support carrier for a 1957 expedition to Antarctica. Since its launch in 1953, the machine has proven to be unbelievably versatile. A modified version is still available today, and it racks up sales around the world.

But it was the personal snowmobile that we all know that really put the Bombardier name on the map.

In the early 1950s, the first small two-stroke engines came on the market. Up until then there was nothing available that would work on a small snowmobile, so as soon as he could, Armand bought one of the new motors. From the outset, it seemed to be exactly what he had wished for. A short time after acquiring it, he set out to build the first small snowmobile, the ancestor of the kind we are familiar with today. The frame of the machine was low to the ground; two skis were installed on the front, and a wide and relatively light tread arrangement was put on the back. In essence, it was a radically scaled-down version of the B7 and B12 — and Bombardier felt it was the ideal machine for carrying one or two people. He tested it for several days in and around his hometown, got it working to his satisfaction, then decided to take it as a gift for his good friend, Father Maurice Ouimet. Ouimet was an Oblate priest who worked with a Native community in Northern Ontario, and he and Bombardier had been friends for years.

However, the reunion had a somewhat unexpected outcome.

This Bombardier snowmobile came out about 1971. It was quite popular at the time, but the machines today are far more advanced.

Bombardier snowmobiles have evolved considerably over the years. This machine, from about 1979, is quite primitive compared to those available today.

Father Ouimet was thrilled to get the prototype snowmobile because he realized that he could use it to visit many of his parishioners who lived far from their nearest community. Those that had homes close by soon heard of the gift Bombardier brought with him and begged to be given the chance to try it out. It is said that during the two days of the visit, the snowmobile was never stopped long enough for it to cool down. One person after another took their turn on this marvellous new machine. By the time Bombardier left for home, he knew his old friend had received a gift he would cherish for a long time.

But that was not the case. One of the members of Father Maurice's parish was a local trapper who made his living trekking by dog sled for many miles through the bush to tend to his traps. These journeys were always long, slow, arduous, and necessary. However, when he saw the little snow machine the Father Ouimet had, the trapper begged his priest to let him use it for tending his trap lines. Father Ouimet gave it to the man, who then stopped using his dogs because he no longer needed them. Instead, the grateful young man worked his trap lines by snowmobile for the next several years. It is said that if ever there was a delighted human being, the trapper was that man. Today, that first snowmobile is in the J. Armand Bombardier Museum.

Back in Valcourt, Bombardier learned what had happened, and the incident reassured him completely. His cherished dream was not just that of inventing a small snow machine. He saw it "as a replacement for the dog sleds used by trappers, prospectors, missionaries, and other persons in the north."[7] In fact, he called it the Ski Dog, for that very reason. The name did not stick, however. Early on, a typographical error in an advertising brochure said "Ski-Doo," and Armand let the mistake stand. He thought the new name was better than the original.

It was the invention of the small Ski-Doo that revolutionized the Bombardier Company. And, while it became a well regarded staple in trapping, it was the sporting and recreational uses of the machine that would make it popular anywhere there was snow. Thousands and thousands of Ski-Doo snowmobiles were made, and their uses were many. In the north, they also "transformed the social life of Inuit and Arctic communities,"[8] and on a personal level, they made Armand Bombardier a wealthy man. The invention now gave him the wherewithal to follow up on a couple of other passions.

Armand loved to sing, in choirs at church, at family or neighbour-hood gatherings, at reunions with former employees, and with his friends. He also was also interested in airplanes and, in due course, bought one. Not surprisingly perhaps, the first thing he did with it was to remove the motor, take it apart, and closely examine the machine in order to see exactly how it worked. Those actions complete, he put the engine back in the plane, and went flying. He was a qualified and competent pilot.

Sadly, however, and as we have indicated earlier, Joseph-Armand Bombardier did not live to witness the empire his company became. He died of cancer in a Sherbrooke hospital on February 18, 1964. He was fifty-six at the time. After his death, family members and others built the company into what it is today: a giant concern involving several branches, active in many countries, and a recognized name in modern public transportation. It is undoubtedly a Canadian success story of epic proportions.

9

NORMAN BREAKEY
Wall-Painter's Dream

Today, few know who Norman James Breakey was. His name is rarely spoken, his picture is not in the papers, his life story has never been the basis of a biography or a television documentary. Even the date of his death is uncertain. However, an invention of his is known around the world and is used by millions. It is commonplace, functional, and affordable. In fact, it has become a vital tool for anyone involved in residential or commercial construction or renovation. Sadly, despite inventing a labour-saving device that has helped so many, doing so brought the man behind the idea little more than disappointment. It certainly never resulted in personal wealth or fame. Nowadays, his story is little more than an obscure footnote in Canadian history. And that should never have been the case. That is because Norman James Breakey invented the paint roller.

The man behind the device came up with it in Toronto in 1940. He may well have been painting a room in his house and grew tired of the thousands of brush strokes the job entailed. He may have been racing against a deadline. He may have despaired at the time it took to cover a wall, to ensure that the tiniest part of it was covered; that the completed work was smooth and not streaked. He may even have been unable to match to his satisfaction an area that had been painted the previous day. Who knows now? Maybe he just had an inspiration and in the midst of his work thought that there had to be a better way. In essence, we have no certain knowledge of what led him to come up with his labour-saving hand tool. We might as well just accept it and be grateful that his original idea led to a new and inventive conclusion. Norman Breakey's idea was a brilliant one and in the many decades since we have greatly benefitted from it.

But do we know anything about the man? The answer is yes, but to a limited and rather contradictory extent. To explain: Norman James Breakey was born on Tuesday, February 25, in Pierson, Manitoba, or Chicago, Illinois. The specific month and day of the week seem to be correct. The year has been recorded as either 1890 or 1891, but probably it is the former that is accurate. It is the year indicated in his military attestation papers that were signed in September 1914.

In completing those same papers, Norman had to answer this question: "In what Town, Township or Parish and in what country were you born?" His written reply, "Chicago, Ill. USA" is both legible and definitive. So, Breakey was an American by birth, despite the often-repeated claim by many that he was born in Canada. However, he did move to, and live in, Manitoba, and that province can rightly share some of the reflected glory attached to his later inventive genius. His only sibling, a sister, Kathleen, was born in Pierson, a small community southwest of Winnipeg. When the future inventor was nine, and his sister scarcely two, their mother died, and the two went to live with an aunt, also in Manitoba. Their father William went to Toronto, likely for work.

Norman stayed on in the western province and as a young man worked in a hardware store in Souris. In the summer of 1913, he joined a militia unit, the 99 Manitoba Dragoons, and was active with that organization for about a year. When the First World War began on August 1, 1914, young Breakey decided to join the army. He took some of the money he had earned as a hardware merchant and bought a train ticket out of town.

Less than a month after the outbreak of hostilities in Europe, he found himself in Valcartier, Quebec, preparing for the physical examination that would lead directly to his becoming a full-time soldier with the Canadian Overseas Expeditionary Force. The date of his acceptance was September 23, 1914. In time, the young man from Manitoba would become an officer, a lieutenant in the army.

In the papers filled out when he took his medical, and then later, on the day he actually became a soldier, there are several items of interest in the forms that were completed. In them, we note a number of things about young Breakey: in fact, probably as many as we learn about anything in his later life.

It was at Valcartier, some eighteen miles north of Quebec City, where the Canadian government established a site for military training. The idea was the brainchild of Major-General Sam Hughes, who, at the time was a confidant of Robert Borden, the prime minister at the time. Hughes, with the blessing of the government of the day, oversaw the expropriation of the farms in the area, and in so doing became highly unpopular with the locals. Nevertheless, he pushed the project through and, within a single month, military buildings, water mains, and even a railway siding had been built on what so recently had been agricultural land. The farm families left, albeit reluctantly, and for the most part blamed Hughes for the disruption of their lives. Then, and later, there was little indication that the general ever cared. His focus was on getting men ready for war. If he stepped on toes and infuriated many, so be it.

The military compound expanded quickly, and, as fast as the new citizen soldiers could be trained and shipped overseas, young, eager, and reasonably able-bodied replacements arrived. Eventually, some 33,000 young Canadians spent time at Valcartier over the next four years. The camp was operational until the Great War ended in 1918, and served as a huge relief camp for hundreds of unemployed and destitute individuals during the Great Depression.

With the outbreak of the Second World War in September 1939, the military moved back. In fact, they are still there. Today, the location is called Base Valcartier and is a far cry for the destination Norman Breakey would have found when he arrived there for his military medical on the second last day of August 1914. He was twenty-four years old at the time.

In what records remain, we know that the future paint-roller inventor was five feet, ten-and-a-half inches tall when he was examined. He was of dark complexion with grey eyes and brown hair. There were other details, many of them rather picayune. For example, according to the records, he had small scars on both his hands and another on the left side of his chin. The medical examiner also noted that the young man had a vaccination mark on his left arm. Surprisingly, his weight was never recorded. Breakey gave his religious affiliation as Methodist, but it is not known how staunch an adherent he might have been to that faith. There is no surviving reference to any church in the years that followed.

Part of the formal medical certificate was pre-printed and attested to by the captain who checked Breakey and stated: "He can see at the required distance with either eye; his heart and lungs are healthy; he has full use of his joints and limbs." One final category might be somewhat surprising to us today, particularly as it required the recruit's self-knowledge alone. He had to declare that he was "not subject to fits of any description." Finally, at the conclusion of the examination, the medical officer had to fill in a blank space on the form before him. Was the recruit "fit" or "unfit?" Breakey was said to be the former. Had he not been, the examiner would have had to indicate in writing, "the cause of unfitness."

Once examined, young Breakey proceeded to complete the usual oath-taking requirements before donning a uniform. At first, he had to indicate that he understood "the terms and nature" of being a soldier. He claimed he did. Then the oath itself required his promise to "be faithful and bear true allegiance to His Majesty King George the Fifth, His Heirs and Successors." Breakey and thousands and thousands of Canadian recruits like him had to affirm that they would "honestly and faithfully defend His Majesty ... against all enemies." In the oath taken, each young soldier promised to not only "obey all orders of His Majesty," but to obey "all the Generals and Officers" they would serve under.

Each individual who took the oath was cautioned, in writing, that if there were any falsehoods in what had been declared, they "would be liable to be punished as provided in the Army Act." There was no indication as to what such punishment might entail — or even if the new soldier gave the warning much thought. Finally, his answers were read back to the applicant. Then, basic training began. When it was completed, Norman Breakey was shipped overseas.

We know little of the man's military record while away from Canada, other than the fact that it must have been successful because he was promoted to the rank of lieutenant. He also met a young woman named Isabella Driver and married her in London, England, in 1916. Subsequently, they had a son, Robert, who did not have much contact with his father. In fact, when the war ended in 1918, Norman apparently left England and returned to Canada, leaving his wife and son behind. Many years later, in 2006, Robert professed an interest in learning more about his father. It is not known if he did so then, or since.

Norman Breakey settled in Toronto and was living there in 1940, the year he came up with the idea for the paint roller. By that time, he would have been fifty years old. Without a doubt, his invention was a wonderful one, but as indicated earlier, it brought him neither wealth nor fame.

Breakey's device was not complicated. "In its simplest form, the paint roller was a cylinder covered with fibrous material and free turning wheels on either end. The axel hooked into a handle, allowing the user to pick up a large quantity of paint on the roller."[1] Breakey continued to work to make sure his invention performed the way he wished. As soon as that point arrived, he knew he needed to apply for a patent. He took all the information he had, all the drawings, all the specifications, and contacted officials at the Patent Office in Ottawa.

In due course, after the investigators in the capital had determined that Breakey's invention was unique and that there was no other like it in Canada or elsewhere, they granted him a patent for the first paint roller in existence. Had there been another, no patent would have been awarded.

It is probable that Breakey was thrilled with his achievement because he was sure his new labour-saving device would surely lead to financial and commercial success. But first he had to sell his idea and locate individuals or companies who would provide sufficient backing so that the new product could be manufactured and sold.

Almost immediately, he encountered a skeptical kind of resistance from potential investors and what would prove to be marked lack of interest. Nevertheless, because he had faith in himself, and perhaps even more faith in his unique discovery, he carried on as best he could. Unfortunately, despite his unshakeable belief that he was on to something revolutionary, he never did find the necessary like-minded individuals with deep pockets who would finance his product. Then, as if that was not enough, he quickly faced a second obstacle: this one more discouraging than the first.

He had to defend his patent, and that took money — and lots of it.

Almost always, as soon as something with presumed potential is invented, the patent for the product is challenged by competing interests — be they individuals or companies. This often happens as a matter of course, and often by competitors who sense financial reward. A similar, but related type of thing occurs if an individual or company appears to tread on another's trademark or brand. For example, if an entrepreneur decided to open a coffee shop and call it "Timothy Hortons," the upstart would undoubtedly receive a lawyer's letter from the well-known restaurant chain. But, while Norman James Breakey had his patent, he could not afford to defend it. That was when Richard Croxton Adams came into the picture.

Adams was an American who claimed to be employed by the well-known Sherwin-Williams paint company. Not long after Breakey was awarded his patent, and still in 1940, the twenty-eight-year-old Adams said he was working in his basement workshop when *he* invented the paint roller. It is believed that he made a slight adaptation to Breakey's roller, applied for a patent, and was successful in getting it. Poor Breakey was left out in the cold, his dream shattered.

Adams does not seem to have ever given the real inventor any acknowledgement, and would always claim the invention was his alone. In fact, in the obituary following his death in La Mesa, California in 1988, there was no hint that the original inventor was anyone other than Adams. "Mr. Adams, an engineer, industrialist and civic leader, held the first patent for the paint roller,"[2] was the way the death listing was worded. The newspaper account noted that Adams was a descendent of two American presidents: John Adams and John Quincy Adams, and that he died of a stroke. One wonders, in passing, what his connection to the country's political hierarchy had to do with the invention. In all likelihood, the paper probably ran the reference because it seemed to be a good story. To be fair however, Richard Croxton Adams *did* have a patent, but it was not the world's first. He did secure the first United States patent on the paint roller. That achievement, given the size of the American market, and the fact that he was able to exploit the product, made Mr. Adams a wealthy man.

The paint roller continues to sell successfully in virtually every country around the world. It is available in the most humble hardware and home-furnishing store, and it is a staple product in every "big box" outlet as well. A cursory inquiry at one of our major dealerships indicated that four thousand paint rollers were purchased there in a single year. This was one store here in Canada. Think of how many rollers passed over the counters in all the places where the product is sold. Norman Breakey would surely be amazed — and saddened — at what might have been for him.

10

CHARLES SAUNDERS
Nourishment for Canada and the World

It has long been acknowledged that wheat is the most valuable cultivated crop in the world. It is also one of the most ancient; originating 20,000 or so years ago. However, exactly where it was first grown is open to speculation, with some historians citing Mesopotamia, while others feel that what we now know as wheat could have descended from the wild grasses found in Europe after the last ice age. The climate at the time was often damp and foggy, and this condition may well have been ideal for such plants to flourish. But even though the precise area is speculative, we do know as fact that wheat was grown and was being traded by our ancient ancestors. Over many centuries, the quality of the crops they grew improved and less desirable grains were removed before cultivation. With population expansion, food was needed and wheat was essential no matter where humans settled.

When the first settlers came to what is now Maritime and central Canada, they did not forget the food preferences and cultivation skills they had known in Britain, France, and elsewhere. As soon as they got here, they raised a few pigs and cattle and fished and hunted, but they also planted crops in order to sustain themselves and their livestock in what was then largely uncharted wilderness. Some items thrived; others did not, but with time and luck, the first small crops of produce came to be. These sprang from the limited quantities of seed the pioneers carried on the ships that brought them.

However, the hardy souls were not coming to a paradise. The new crops sometimes yielded little. Harsh winters and scorching summers hampered growth, and often plant diseases and insect infestations ruined results. Yet, these farmers carried on, determined to succeed in the places they now called home. They dealt with misfortune, cleared more land, and cultivated ever more extensive holdings. Along with the manufactured goods they had to order from England and elsewhere, they also imported more seed grains, and wheat was the major one. Some varieties that might have thrived in England or Scotland did poorly here, while others grew well. However, none of the results came about without long hours of labour and an abundance of patience and hope.

Every farmer tending his crops wanted greater yields, and when the weather co-operated, he got them. Initially, he ploughed his fields with oxen, then with horses, and finally with tractors that over time became absolute necessities. Today, the giant implements used in agriculture were not even dreams for the settlers of the past. But those early farmers laid the foundation for the vast and varied systems we have now. They tilled the soil, planted the seed, and harvested the results. So do the men and women who work the land today.

But it was not just the implements that improved. The quality of the seed that was planted gradually got better. Often a farmer who obtained a high yield in his crops sold some of his best seed to his neighbours. This was most often wheat, largely because it was so versatile and necessary. The good grains had to be cleaned before a follow-up planting was done, however. Crops of weeds were of no help to anyone, so rudimentary machines such as fanning mills helped separate the weed kernels from the good seed.

Sometimes the grain that was brought to Canada was of a particular strain. It was planted and the quality of the harvest was assessed. If the crop was good, then more of the same type of seed was planted. Gradually, the best varieties spread and became widely known — not only in Canada, but in the crop-growing areas of the United States and elsewhere. Good seed was traded internationally and it became one of the staples that fed the ever-expanding populace.

And it was this population growth that in time led to the opening up of the west. Thousands of acres in what is now Manitoba, Saskatchewan, and Alberta were largely empty; the soil unbroken. A succession of governments in the east championed the idea of westward expansion and, over time, the movement grew. In fact, "between 1896 and 1914, more than two million settlers from Europe and the United States poured into the prairies in the greatest wave of immigration in Canadian history."[1]

Those intrepid souls who made the trek west often faced hardship so severe that they sometimes gave up and fled their hardscrabble existence on the plains. But the majority who reached the prairies stayed there. They struggled, certainly, but they persisted. Tracts of land were cultivated, various grains were planted and, as in the east, farmers sought the best yields. Early on, a hard, red variety of wheat looked to be the most promising. It probably originated in the Ukraine and while there are several legends to that effect, all of them are speculative. What is known is that a man named David Fife somehow obtained a few kernels of the strain and he planted them on his farm near Norwood, Ontario. This was about 1842. The grains grew well, so Fife planted more of them. Inside a couple of years, his crops were the best in the area, and the seed he was cultivating was given his name.

It then took about two decades to happen, but by the mid-1860s, Fife wheat, or Red Fife as it was commonly called, had become popular, not just in Ontario, but also in the west. It became Canada's wheat and remained so for over four decades. Then, enterprising members of one scientifically inclined family took the improvement of grain to an entirely new level. The family's name was Saunders and their hometown was London, Ontario. They have been and always will be associated with the development of seed grains and the refinement of agricultural practices in Canada, and to a lesser extent in the United States, as well. We owe them much.

The Canadian patriarch was English-born William Saunders, who came to London in 1848. He became a chemist there and opened a drugstore, but bought a small farm a little later on so that he could conduct scientific experimentation on field crops. This work brought him much success and when the first of a series of experimental farms was established by the then Dominion Department of Agriculture, Saunders was named director of the branch. He and his wife Sarah had one daughter and five sons, and the boys shared many of the father's interests. In fact, one in particular became known in Manitoba for his cross-breeding of Red Fife wheat.

Percy Saunders was the second-youngest son. With his brothers, he always called London his real home and, like them, lived above the family drugstore on Dundas Street in the heart of the city. When his father bought the farm, however, young Percy loved the rural atmosphere and, while there, made several rudimentary attempts at cultivating better crops of field grains. He went on to graduate from the University of Toronto and soon afterwards found work at the newly established experimental farm in Agassiz, Manitoba. While there, about 1892, he became familiar with an Indian wheat strain called Hard Red Calcutta. It matured early, although not always before the first frosts blanketed the prairies. As young Saunders was quite familiar with Red Fife, he took it upon himself to see what kind of grain would result if Red Fife and the Red Calcutta wheat were crossed. The whole idea was to somehow produce ever-earlier maturation times in order to beat the yearly freeze-up across the west.

As far as Percy Saunders was concerned, the cross-breeding was a success, but not completely so. It would be left to another Saunders brother to come up with an early-maturing breakthrough. This family member was Charles, and the bulk of his scientific work was done at the most important experimental farm of all, the central location in Ottawa. That fact that he was there came at the end of a rather extensive personal odyssey, one that was neither his wish nor his intent, but it happened because of circumstances that were beyond his control. Fortunately, for Canada and the world, his presence and work in the nation's capital was fortuitous indeed. But, first a few words about the man himself.

Charles Saunders was seventy when he died, but he was never physically strong. And while he loved being outdoors, particularly if he could study and experiment with plants, he had a natural affinity for music. Both parents possessed the same love, although it was his mother who seems to have had the greater impact. When Charles was only fifteen, he performed at a concert at Victoria Hall in London. On that occasion, he played a flute solo, although he was equally talented on the violin. His mother and two of his brothers were on the stage that night, as well, and the mother sang.

Saskatchewan Archives Board R-A350.

Sir Charles Saunders was born in London, Ontario, but moved to Ottawa, where he discovered Marquis, a strain of hard wheat that grew well in the Canadian Prairies.

After he completed secondary school, Charles went on to the University of Toronto, and later earned his PhD at Johns Hopkins in Baltimore. After graduation from there, he spent one year as a chemistry professor at the University of Kentucky. By 1893, at the age of twenty-six, he had married, was writing a column about music for a magazine and, with his wife, had opened a music school in Toronto. In the meantime, his parents had moved to Ottawa because the father needed to be there to oversee the experimental farm network.

The young teacher struggled to make the music business pay, but never succeeded in doing so. He had students, some of whom were quite enthusiastic, but there were not enough of them. He and his wife Mary cut corners and barely managed to pay the rent on the facilities for their fledgling business, but only for a time. To make matters worse, Charles's health gradually deteriorated and he found that constant money worries added to his discomfort. Finally, he and his wife pulled up stakes in Toronto and moved to Ottawa because his father had him appointed to a position as "cerealist" at the experimental farm. Even though the son initially balked at taking the job, his financial predicament forced him to put aside whatever scruples he might have harboured about doing so.

He was ideally suited for the role. In fact, this "sick man who couldn't make a living as a musician in Toronto, added more wealth to his country than anyone before or since."[2] He did it through sheer perseverance, absolute dedication to the task, and remarkable foresight. There were, however, many steps, and much trial and error before Charles Saunders could call what he did a success. He spent not only days, but weeks and months planting varieties of seed wheat, tending them carefully, then noting with precision the yield from each. Some turned out to be worthless, others showed some promise; a few surpassed the rest.

One day, Doctor Saunders happened to note a single head of wheat that seemed far superior to *all* the others. He painstakingly planted each seed from the one head and, to his immense relief, saw twelve plants appear the following spring. The strain was a pure and more resilient cross of Red Fife and Hard Red Calcutta, essentially the combination his brother came up with in 1892. Now, however, Charles Saunders had both the time and the dedication to work towards the further development of the cross. He took the "new and improved" seeds, carefully cultivated them and, by 1907, had enough to send a small sample of the grains to Indian Head, Saskatchewan, for testing at an experimental farm there.

The new wheat was given the name Marquis.

The variety proved itself within a single growing season. More samples were harvested and were distributed to several farms across the west. The strain grew well in all locations, matured earlier than any before it, and by 1911, "in the United States, as well as in Canada, and in practically all wheat growing areas of the world, Marquis became known as the 'yardstick of quality.'" In fact, "in Canada alone, the achievement of Marquis was phenomenal and of great importance in the development of western and northwestern Canada."[3]

But being able to relax was not part of Charles Saunders's temperament. He was never complacent about what he had achieved, and continued his daily research. He turned his attention to the milling qualities of the new variety, then to the study of what the new wheat was like when made into bread. In the latter, he was particularly thrilled with the results. Bread made from Marquis was tasty, nourishing, and of excellent quality. All three factors were important in the consideration of the wheat as a domestic and international product. Within a few years, it was the wheat that every farmer preferred to plant. The fact that it matured up to ten days earlier than other kinds set it apart for years afterwards. That was why, "by 1920 over 90% of the 17 million acres of wheat in western Canada were Marquis. The introduction of this hardy, early-ripening wheat allowed farmers to grow the crops further north, doubling the amount of arable land on the prairies, and was responsible for the reputation Canada gained for producing the best hard spring wheat in the world."[4]

Charles Saunders continued his work in Ottawa, experimenting with and growing other grains such as oats and barley. He also travelled extensively and lectured widely to academic and other audiences about the discovery and development of Marquis. No matter what the month, he was in such popular demand that he often had to pass up invitations to speak at agricultural societies and universities. Yet he never wavered in his desire to educate everyone he could about the excellence of the wheat and its worth to Canada and the world.

Not surprisingly, his work was soon widely known, as was he. He was awarded honourary degrees, medals, plaques, certificates, and civic citations. A new wheat variety that originated at the Ottawa Experimental Farm was named after him. But no matter what the honour might have been, it did nothing to improve his health. Ultimately, he could not even walk over rough farm fields without suffering excruciating pain. When he finally could not continue to work as he wished, he was forced to leave his workplace behind. He resigned his Ottawa position.

Yet Doctor Saunders could not sit idle. After a period of rest in the nation's capital, he and Mary decided to go to France for a time. They admired French culture and found the language of particular interest. Both had been fluent in French for several decades, and, because of that, decided to enroll at the Sorbonne in order to become completely immersed in everything that the renowned university could offer. They stayed over two years, and, fortunately, the change in scenery proved beneficial. Charles's health improved and the couple returned to Canada.

Despite the fact that he had been out of this country, and was retired from the experimental farm research job, his achievements there had not been forgotten. In fact, the flow of honours continued, with more medals, degrees, and fellowships. Then in 1934, came the ultimate honour: His Majesty, King George V made Charles a knight. By this time, Sir Charles and Lady Saunders were living in Toronto.

Unfortunately, they still had money worries.

When Charles had to leave his experimental farm employment, he was "entitled to a pension of but $900.00 a year."[5] Even for the time, the amount was miserly, so admirers with influence petitioned the government to grant him an increase. A small sum was allotted, but it was left to the farmers of Canada to agitate for more. Ultimately, their voices were heard, and this quiet, patient, and brilliant man was granted a sum of $5000.00 a year.

Sir Charles and Lady Saunders lived quietly in retirement, and continued to explore the subtleties of the French language. Sir Charles even wrote a book on the subject. *Essais et Vers* was published in both Montreal and New York, and it was praised in those cities and in France. Saunders welcomed the acclaim, perhaps as much as he was grateful for the recognition he received for his Marquis wheat discovery. He passed away on July 25, 1935, and his wife followed, a year to the day later. Sadly, too few know much about his accomplishments today.

11

JACQUES PLANTE
Facing the World's Fastest Game

For some time now, much attention has been paid to the problem of concussions in hockey, particularly because some of the greatest players have been sidelined with head injuries. Foremost among them has been Sidney Crosby, the Nova Scotia superstar who, at the tender age of nineteen, became captain of the National Hockey League's Pittsburgh Penguins. Crosby was on the receiving end of a vicious shoulder-to-the-head check from Dave Steckel of the Washington Capitals during the outdoor Winter Classic in Pittsburgh on New Year's Day, 2011. The check caused the young Penguin captain to lurch to his side, lose his balance, and crumple to the ice. He lay there in agony, then got to his feet and with his head down, slowly made his way towards the bench. By this time, the club trainer was on the ice, doing what he could to assist the Pittsburgh star. No penalty was called on the play, but Crosby was obviously badly hurt.

The Crosby story is relatively recent, and the account of the matter was highly publicized at the time and since. Contrast this with injuries of another kind, during another era when our national sport was arguably even more dangerous than it is today. Back then, there were injuries — lots of them — but relatively few received sustained attention. Often the most serious were suffered by goaltenders, for a rather obvious reason. Until about 1960, the player in the net risked his life every time he played. A single, bullet-like slap shot to the face could have been fatal because the goalies of the day didn't wear masks. It just wasn't done at the time.

The injuries were commonplace; they happened to virtually every netminder at every level. One of these players often on the receiving end of a puck to the head was a highly skilled professional goaltender named Jacques Plante. He played more than eight hundred games in the National Hockey League, and was hurt many, many times. In the first third of his career alone, he had received over two hundred facial stitches. As well, his nose had been broken four times, his jaw had been shattered, both his cheekbones had been broken, and he had suffered at least one concussion. He sometimes missed entire games with injuries, but more often than not, after his face was cut, he would go to the dressing room, get stitched up, and return to the ice. Then he had to face more shots that came at him with flying precision, knowing all the while that if another hit him in the face, he would have to be patched up again, and there would be stitches on top of stitches. No wonder observers said he was beginning to look like a Frankenstein monster: he had cuts, scars, and bruises from chin to forehead and from ear to ear. Each reflected a particular incident. All were painful when they happened.

And there were lots of these.

One of these injuries occurred at Madison Square Gardens in New York City on November 1, 1959. Plante was in goal for the Montreal Canadiens, when Andy Bathgate of the hometown Rangers hit him on the left side of his nose with a drive from fifteen feet out. The shot was a scorcher, and Plante neither saw nor was prepared for it. He was knocked cold, and went down, bleeding profusely. Habs trainer Hector Dubois rushed to Plante's aid, assisted as he regained consciousness, and attempted to staunch the blood flow with a towel. A few minutes later, the injured netminder received several stitches to close the ugly three-inch gash to his face. Then, after a heated confrontation with his coach, Plante returned to the game.

At the time, the coach of the Canadiens was Toe Blake, a stern task-master who insisted on complete obedience from his players. The confrontation itself occurred because in the weeks leading up to that night in New York, Jacques Plante made no secret of the fact that he believed every goaltender should wear a mask for protection from injury. He even had a prototype made, and had worn it in a few practices. Blake grudgingly permitted that, but was dead set against his goalie using the mask in any regular game. This, the coach felt, was because if his netminder was wearing a mask during a game that was lost, everyone would blame the mask. And to Blake, every game his teams ever played was important.

On the night in question, however, Blake had a particular problem. He was acutely aware of the seriousness of Plante's injury, but felt he had to get his netminder back in the goal, because there was no other qualified goaltender he could use. There was a so-called "house goalie," a young Madison Square Gardens usher who sometimes practised with the Rangers, but he had never played pro. At the time, no backup goaltender was carried by any team in the National Hockey League.

Blake assessed the situation and talked to Plante about it. Initially, neither man mentioned the mask, but when he saw how seriously his goaltender was hurt, Blake felt he had to broach the subject. He swallowed his pride and gave his permission. "Plante reached down into the bag beside him and grabbed for his mask. He gently placed it over his battered, swollen face, and slowly began walking up the flight of six stairs that led to the ice and towards his place in history."[1] On that memorable night, for the first time ever, a goaltender in the National Hockey League played in a regular game and wore a face mask because he needed to do so.

To Plante, wearing that mask was a victory of sorts. In time, too, it became a victory for every goaltender, everywhere. Anyone who follows hockey is well aware of the fact that today a face mask is an integral part of the netminder's equipment. In fact, if the goalie's mask is knocked off at any time during a game, play is immediately whistled dead. Most fans now can scarcely believe that at one time, no goaltender wore a mask at all. In fact, playing without one today would be looked upon as sheer lunacy. But the struggle to get to this point was controversial and long, and Jacques Plante was the central figure who brought mandatory mask-wearing to the fore.

Photo courtesy of the author.

It is not even necessary to enter the Hockey Hall of Fame to see a tribute to Jacques Plante. An early face mask and his picture are displayed outside the Hall.

There were others before Plante who tried masks, but there were few of them, and the use of it was sporadic. As well, early versions were makeshift, uncomfortable, or merely adaptations of facial protection used in another sport. The earliest-known wearer was a man named Clint Benedict. He played 362 games in the NHL for Ottawa, and later Montreal, in the years immediately prior to the Great Depression. His mask was a wire-leather contraption and he only wore it for about half a dozen games. He put it aside because he said it obscured his vision.

At least one woman was also a pioneer in the use of the mask. Elizabeth Graham, who played goal for Queen's University in Kingston, wore a fencing mask in a game in Montreal in 1927. The rather sexist male sportswriters of the day claimed she did it because she had a beautiful face and did not want it disfigured by a flying puck. Perhaps as well, she did not want to lose any teeth in the game. One or two others tried a mask on occasion, but without doubt, it was Jacques Plante who pioneered the equipment — and he had an illustrious career before and after its adaptation.

Jacques Plante was born on a farm near Shawinigan, Quebec. The eldest of eleven children in a home where money was scarce, he had an innate drive to make something of himself. He loved hockey, but because he suffered from asthma, found he could not keep up with other skaters in pickup games. For that reason, he elected to play in the net, and in that position would, as an adult, become "one of the most influential goaltenders in NHL history."[2] All this started when he was about twelve, stopping pucks when he was generally the youngest player on the team. As a youngster, he played as often as he could, and developed his on-ice skills accordingly. He made his professional debut with Quebec Citadelles in 1947, shortly after graduating from high school. Four years later, he was between the pipes for Montreal Canadiens.

Plante had a long, illustrious career with Montreal before being traded to New York. Ironically, he then became a teammate with Andy Bathgate, the shooter most responsible for the introduction of the mask. Plante later played for St. Louis, Toronto, and briefly, Boston, as well. He even backstopped for Edmonton Oilers before they became part of the National Hockey League. Over the years, he retired twice, then returned to the game. After three years of retirement the first time, he got talked into signing with St. Louis Blues. There, he and another goaltender named Glenn Hall won the Vezina Trophy, awarded to the best goalies in the league. Plante had won it six times earlier, a record that still stands.

As we have mentioned above, Jacques Plante was bothered by asthma as a child. Unfortunately, he never really grew out of it, and the condition was an ongoing irritant for the rest of his life. He missed time because he couldn't breathe, had to be pulled from games, and even collapsed during play because he couldn't breathe. Sometimes, the asthma caused dizzy spells. At other times, it weakened him, or made him nauseous. Occasionally, even the atmosphere in some rinks made the problem worse. This was particularly bothersome in Toronto, and when he played for the Maple Leafs he adjusted his living location so that he could, in so far as possible, get away from some of the excess humidity of southern Ontario. Nevertheless, Plante played, and played well, setting long-standing records, and inculcating a style of netminding that was as revolutionary as the mask he eventually got to wear.

For example, he skated away from his goal.

Today, such a practice is commonplace, an integral part of the game, but not until Plante set the standard. He was really the first goaltender who regularly went behind his own net to retrieve the puck. As well, he often skated away from the crease, stickhandled as he went, then passed to one of his teammates. These moves were not only new and unexpected; they were the bane of opponents, and they often gave the Canadiens a jump on whomever they faced. Prior to Plante, goaltenders simply stayed in the net, and at most, deflected the puck to either the side, or by luck, to one of their own skaters. When questioned about this action, Plante used to say that as long as he had the puck in his possession; opponents did not have it in theirs.

But it is the mask that people remember most when they think of Plante, and it did not come about overnight. He had used a makeshift face protector for practice purposes some three years before he was permitted to use it in that memorable game in New York.

And the precedent came about because of earlier injuries that did not even happen in games — they occurred during practices, and were caused by his own teammates! "In 1954 his right cheekbone was fractured by a ... shot that sidelined him for five weeks. The following year his left cheekbone, and nose, were fractured by one shot."[3] After the second injury, he tried a makeshift mask a fan sent him, but like Clint Benedict before him, felt he could not follow the puck as well. The thing interfered with his vision.

Plante was intrigued though, but still not convinced. "I wore it for a while," he recalled. Then he added, "I was just thinking of taking it off when a puck hit the mask right in front of the eyeholes. I kept it on religiously in practice from then on, wondering all the while about what kind of a mask would be practical for wearing in games."[4]

It was not until the spring of 1958 that the idea of a mask evolved further. Again Plante was badly cut, this time by a shot that hit him flush in the forehead. Again it was a fan that broached the idea of, not only another mask, but this time one made out of fibreglass that could be moulded to Plante's face. The fan even wrote to Plante, explaining the suggestion at length. It took another year, but in that time, the Habs goaltender started to believe the letter-writer might be right. One day, he agreed to go with a couple of team medical people to Montreal General Hospital where a plaster impression was made of his face. Finally, by using the cast, and a combination of small, woven rope soaked in fibreglass resin, a face mask was especially fashioned for Plante. The thing looked ghastly, but it fit reasonably well. And, once strips of sponge were glued to the inside, the prototype was fairly comfortable to wear. Plante tried it, had various adjustments made, and declared himself satisfied with the invention. Later on, he had it modified several times, and each adjustment came closer to what he felt should be a mandatory piece of goaltender equipment. He even started a company to make masks, his own included.

As a point of interest regarding those first exploratory masks, hockey fans who visit the Hockey Hall of Fame in Toronto can see for themselves how rudimentary the early masks actually were. The hall has an extensive collection, from the earliest concoctions up to the types of face protectors used today. Just seeing the exhibit leaves the viewer with a greater appreciation of just how much Jacques Plante gave to the game.

But of course, the man should be remembered for more than the mask. His innovative style brought fans to his games; but so did the play-making skills that led to the records he set. Many of these still stand, and may continue to do so for the foreseeable future. Most observers of the sport doubt that Plante's Vezina Trophy record will ever be broken. Today, there are many good goaltenders, but the number of games each plays is much less than it used to be. Every team has more than one netminder now, and should the main one be injured, he would be pulled from the game more quickly than in Jacques Plante's day. Being stitched up after a serious facial cut, then immediately returning to the net does not happen today. There are also aspects of the Plante record that likely will not be replicated. For example: retiring for three years, then returning to the game and winning the Vezina Trophy after such a layoff would now seem to be problematic at best.

But what about this man behind the mask?

Jacques Plante was unique, probably as much so as any player who ever laced a pair of skates. He was highly intelligent, inquisitive, and firm in his belief that he could play goal as well as anyone ever did. In this, he has not been supplanted. He overcame health problems to stay competitive, and he disregarded the opinions of detractors who said he was being

Photo courtesy of the author.

Two of the individuals whose work is described in this book are forever enshrined in the Hockey Hall of Fame. Both Tim Horton and Jacques Plante are well remembered by followers of Canada's most popular sport.

cowardly by opting to wear a mask in goal. All these so-called "purists" became more muted in their remarks when Plante won games while masked. Early on, he backstopped the Canadiens to ten straight victories immediately after donning the mask on that memorable night in New York, when Bathgate's shot did so much damage.

Jacques Plante was selected to play on eight NHL all-star teams and later, had his #1 sweater permanently retired by the Canadiens. Sadly, during the winter of 1986, Plante became seriously ill when he and his wife were on a visit to Switzerland. He was taken to a Geneva hospital, but never left. The great goaltender, the father of the face mask, and the member of the Hockey Hall of Fame, died of cancer on February 27 that year. He was fifty-seven years old. He is buried in Switzerland.

12

WILBUR FRANKS
Safe Suiting for Flying

There have been lots of unsung heroes in Canada, and Wilbur Rounding Franks was one of them. He was a quiet, unassuming, dedicated individual who for many years was a medical researcher and university teacher. He died in 1986, but his work lives on, especially for those who fly high-performance aircraft, or leave the earth on rocket ships. And, as is often the case in this country, few know of the man, and fewer are aware of his greatest accomplishment. Even those who are kept from danger because of his invention generally are unaware of the person behind it. Perhaps he would have wanted that; perhaps not. Yet Franks made the workplace safer for a highly select group of professionals. In fact, many owe their lives to him.

The public part of the Franks story goes back to the mid-1920s, when he studied at the University of Toronto, graduated from the same school, then began to work there in the field of medicine. He was a brilliant scholar, then a hard-working and well-liked professor who conducted cancer research at a time when that dreaded illness meant almost certain death. Franks was absolutely determined to ameliorate the pain of the disease and, in the best of all possible worlds, to find a cure for it. In those years that led up to the Second World War, he remained at his alma mater and continued with his research there. During this time, he met the famous, Nobel Prize–winning Doctor Frederick Banting. The two became friends and shared many of the same hopes, dreams — and fears.

It was the latter that led them to work together on wartime projects for the Allied cause and, in Franks's case, for Britain's Royal Air Force in particular. Such activities were a far cry from Banting's insulin discovery or Franks's work on cancer, but there was a need that meant much to both individuals. The need stemmed mainly from Banting's fears about the resurgence of Germany under the leadership of Adolf Hitler. The Nobel winner warned the government of the day that Hitler was not to be trusted and that Canada was woefully unprepared for the war he was certain was coming. He found a like mind in Franks, and both resolved to do all they could to assist the Allied cause. That was why both felt that the University of Toronto had the facilities, the personnel, and the obligation to stand up to the machinations of the German tyrant. In fact, Banting staked his reputation on the matter, and his persistence paid off.

The Nobel Laureate's great strength "was his ability to mobilize the talents of others,"[1] and he did so in this case. A top-secret facility, initially called the Number 1 Clinical Investigation Unit, and later the RCAF Institute of Aviation Medicine, was established near the intersection of Avenue Road and Eglinton Avenue in Toronto. Banting was in charge, and he co-opted a host of renowned experts to work there. Wilbur Franks was among them, and he was assigned an unusual task. He was to find some way to make flying safer in the new fighter planes that were coming on stream in Britain, specifically the Hurricane and the Spitfire. Both were fast, manoeuvrable and deadly — sometimes for the pilots who flew them, and for a reason that was both predictable, unintentional, certainly, but unavoidable in some extremely trying circumstances.

Up until that time, fighter planes were relatively slow-moving, although during dives, they obviously sped up. And when that happened, vastly increased stress and pressure on the pilot occurred, particularly when the flyer pulled out of a steep dive. In such a circumstance, blood from the pilot's upper body was then forced downwards into his lower trunk, legs, and feet. The increased gravitational, or G-force as it was called, was the reason for the blood movement.

Unfortunately, because of the downward pressure on the blood, it moved away from the flyer's head and brain and affected the individual's capacity to function properly, to see, and to think. If enough blood flowed away from the person's head, the subject could — and sometime did — lose consciousness. This state of affairs had already been noted during aerial battles, or dogfights, as they were called in the First World War. Sometimes flyers dove toward a foe, intending to shoot the enemy down, but the fighters would end up crashing when they went to pull out of the dive. "As early as 1917, there were documented cases of pilots' loss of consciousness due to G-forces that were referred to as 'fainting in the air.'"[2]

During the time between wars, aircraft advancement had continued and with the development of newer and faster planes, the incidence of pilot problems increased substantially. The new fighter planes could dive, turn, climb, and attack quicker and with more dexterity than any of their predecessors. Correspondingly, G-force pressure was increasing, too, and suddenly more and more flyers started reporting light-headedness, vision impairment, and even instances where they were sure they had suffered brief periods of unconsciousness or blackouts in flight.

Because they faced this state of affairs, Royal Air Force senior commanders began to demand that plane builders, equipment manufacturers, and the scientific community devise some way of protecting pilots. Too many were being lost too quickly, to G-force problems that led to crashes. It was bad enough to have a plane shot down; it was worse to have it crash for no obvious reason. That was why G-force was sometimes referred to as the "invisible enemy."[3]

Today, we are unaware of Wilbur Franks' first thoughts when Frederick Banting handed him the problem. At the time, Franks had never flown at all, in any kind of aircraft, so something that caused concern in fighter planes may not initially have been much more than an intellectual exercise for him. He remembered, however, that he had faced a related kind of situation during his years as a scientist. This had to do with the breaking of test tubes when they had to be spun in a centrifuge during various studies. Back then, he had figured out a rather ingenious method of preventing such breakage. The idea involved water.

Franks solved the annoying breakage problem when he placed each test tube in a larger container of water. When the test tube had to be spun in a centrifuge, the water shield eliminated any breaking. He wondered if a pilot in a spinning or diving plane could somehow be immersed in water in such a way that it would protect him.

Initially, his colleagues thought that even the idea was preposterous. They scoffed at Franks whenever he mentioned his solution, but he refused to toss his idea away. Instead, he hired a tailor and had the man make a suit, not out of cloth, but of rubber. Once the suit was finished, he instructed the maker to make leakproof pouches that could be attached to it. These were placed around the lower legs, thighs, midsection, and chest. With the tailor's help, Franks pulled the thing on, had the pouches filled with water, and when that was done, called the garment his anti-G suit. It was the first one ever. Now, he intended to try it out as soon as possible.

A few days later, on a warm, pleasant morning in May 1940, Franks and a couple of others piled into a vehicle and drove north from Toronto to the military airport at Camp Borden, near Barrie, Ontario. There, in a small aircraft called a Fleet Finch, Franks flew for the first time in his life. He did so wearing the rubber suit with the water-filled pouches attached. He needed assistance getting the garment on and stood as it was pulled up around him. And even though the pilot probably wondered at the time about what in the world his passenger was wearing — and why — there is no record today of what might have been said. We do know though, that the man who took the controls that day was instructed to do plenty of tight turns, dives, and assorted aerobatics. He was more than happy to do so.

Franks was helped on board and he and the pilot took their seats. Five minutes later, they were in the air, and heading away from their starting point. Gradually, the plane climbed higher and higher, to a safe level for aerobatics. Then, once he reached this altitude, the pilot put the Finch into a steep dive. It careened earthward, gained speed, and vibrated in the rushing wind. At the bottom of the dip, the pilot pulled up sharply, while his passenger held on. Franks's first experience in the air was like a wild roller coaster ride.

There were more dives, abrupt directional changes, and over-the-top moves that taxed the airframe of the plane and its occupants. Through it all, Franks neither lost his nerve nor his breakfast, even though on one or two of the dives both men were exposed to quite high gravitational pulls. In fact, "while pulling out of a steep dive; the pilot experienced a temporary blackout, but Franks did not." That was because of the rubber suit he wore, and the water pouches that surrounded and protected him from the violent lurches that would be involved in aerial combat. However, even though "he was jubilant that his concept worked … it had not been a pleasant experience."[4] The problem lay in the cut of the garment he wore. Measurements for the suit had been made while he was standing, and he and the others had put it on him as he stood. But then because he had to be seated during his flight, the tight-fitting rubber suit became extremely uncomfortable. "When the pressure hit, I thought it was going to cut me in two,"[5] he admitted later on.

Nevertheless, the suit worked. Now it had to be refined in order to make it more adaptable for the needs of pilots. To do that meant it had to be made less cumbersome, easier to get into, possibly lighter, but still reliable. And right away, Franks realized that it did not have to come up to the chest; he felt it would suffice if it covered only the legs and the abdomen. Those were the areas of the body where blood pooled — not in the chest.

Because of the flight, Franks gained important knowledge of the impact of increased G-forces. He knew, as did everyone else, what a single G meant. It is the force that holds us on to the earth: the strength of gravity. He now had experience of what happened when that force was increased: to 2 Gs, to 3 Gs, and more. At the time, it was known that the greater the G-force; the more certain the flyer would encounter problems in the air. That is the case today, as it was the morning Franks flew.

Today our Air Force refers to a number of sensations that can be noted while flying in situations where there are various stages of G-force. At 2 G, a pilot experiences "downward pressure, heaviness of limbs and head, and movement is difficult." At 3 G, there is "extreme heaviness of limbs and body." Anyone experiencing "from 3 to 4 G, would note a dimming of vision, or grey-out." Then, from "3.5 to 4.5, loss of peripheral vision," and from "4 G to 5.5, total loss of vision, or blackout." Additional Gs beyond this would lead to loss of consciousness after about five seconds.[6]

Because that first G suit was so uncomfortable in flight, Franks looked for as many ways as he could to modify it. Not having it come up to the chest was the most obvious and easy change. But then he started checking other areas, as well, in order to make it more pleasant to wear while sitting. Here, he removed the water pads from the back and front of the knees, and from the groin area. When that was done, a pilot wearing the suit would be able to sit and hopefully be able to concentrate on flying, not on what was initially regarded as just another safety gimmick. The tailor was brought back and after a couple of weeks of trial and error, he produced a new suit. This one would be tested, not by the inventor himself, but by an actual pilot.

A Royal Air Force flyer who was in Canada at the time donned the new G suit and wore it during evaluation flights from Malton, where Toronto's Pearson International Airport is today. He did it in a Spitfire; the amazing new aircraft that many believed was the key to winning the Battle of Britain. The plane was brought to Canada for one reason: to test the G suit. At the time, similar flights in the skies over England would have been extremely risky. There were just too many German marauders waiting to shoot down any Allied plane they saw.

There were three test flights in Canada and the RAF officer at the controls wore the suit during all of them. He put the plane through its paces: plenty of loops, dives, high-speed turns, and sharp acceleration. All of these moves did not just imitate the kind of thing expected in an air fight; they enabled the pilot to experience the G suit and to determine the type of protection it would give in a skirmish against a skilled enemy. By the time he had landed following the third flight, he had pretty well made up his mind as to the suit's effectiveness. Despite some reservations, he endorsed it.

But now, another element of the times had to be considered. Canada was at war, and a Spitfire — any Spitfire — operating in Canadian skies was going to be noticed. Why such an advanced aircraft was there at all would be reason for speculation. Also, why was it doing test flights? What was being tested? Why was there secrecy around every flight? All of these factors had to be considered because no one knew for sure if German spies might have been watching and reporting their observations to Berlin. For those reasons, all the in-air work was stopped. Instead, the G suit design team realized they had to find another way. To that end, they built a large centrifuge — based roughly on the type of machines that Franks had used when he was spinning those test tubes years earlier. Now though, the centrifuge would carry a human being.

In order to finance the construction project, the Institute of Aviation Medicine secured substantial grant money from the National Research Council in Ottawa. Then, with the funds assured, a large space in one of the Toronto buildings was made available for the placement of the apparatus. As war raged overseas, the need for the machine was pressing, and its construction was carried out in haste. Once complete, it was the first such centrifuge in any Allied country. "The Germans had built a smaller, less sophisticated version before the war, but Franks' device was the first that could mimic the effects of aircraft acceleration on the human body."[7]

The centrifuge, when fully operational "consisted of a cockpit swung at the end of a rotating horizontal arm. The guinea-pig airman, racing around the circle thirty-two feet in diameter, experienced exactly the same sensations — blindness, followed by a complete blackout — that the fighter pilot felt in diving his Spitfire or forcing it into tight turns."[8] So now, because of the war, neither Franks nor his associates took the time to revel in what they had accomplished. Several young pilots, all wearing G suits, were spun in the centrifuge, and none reported serious or insurmountable problems. Some "washed out," though, and realized they were not suited to high-speed air combat.

With the early testing of the suit complete, additional ones were made, modified and tried, and, in due course, mass production of the garments began. They were first used in combat circumstances by Royal Air Force pilots during the amphibious landings at Oran, Morocco, in November 1942. Fortunately for the Allied cause, they alone had G suits to wear. No German pilot was similarly equipped, and, without them, they faced additional danger in the air. As a kind of acknowledgement of his ingenuity, Wilbur Franks received the Order of the British Empire two years later. The citation for the OBE said, in part, that his invention of the G suit had given "the Allied Forces a tremendous tactical advantage," and that it had saved the lives of "thousands of Allied fighter pilots."[9]

In the early stages of G suit design, water pouches were always used to provide the necessary flyer protection. But the water was heavy, cumbersome, and somewhat uncomfortable for the pilot. For that reason, the design team began to consider using air in the G-suit bladders instead of water. They correctly reasoned that air would eliminate the negative elements of water. However, all the older planes during the

Second World War and earlier lacked the engine power needed to pro-duce sufficient air pressure to fill the suit compartments. That was not the case when jet planes came along.

With the coming of more advanced aircraft, no G suits used water, nor do they today. All modern fighter planes have plenty of power to inflate the suits with air and can do so during flights. Now the inflation is automatic, and only occurs as needed.

Some time ago, the author flew in the rear seat of one of Canada's F-18 fighter jets. During an hour-and-twenty-minute training exercise, I experienced first-hand the weight of increased gravitational pressure and how the modern G suit compensates for it.

When we climbed aboard the plane, both the pilot and I wore unin-flated G suits. They were just part of the necessary equipment, like the helmets on our heads and the flight boots on our feet. The suits were not uncomfortable at all. In fact, they were barely noticeable during the taxi out to the runway, or during the initial moments of straight-and-level flight. It was only when the pilot initiated tight turns, fast climbs, or the pull-out from dives that I understood how the G suit worked and how vital it is.

There were three other fighters airborne when I flew, and the pilots carried out simulated dogfights at about the 20,000-foot altitude. There were plenty of times when we were upside down, involved in abrupt directional changes, climbing quickly, or diving. My G suit would inflate as required, and I was aware of it doing so. However, when the pilot elected to show me how insidious G force can be, he suggested I watch one of the three digital screens on the console in front of me. That was where the number of Gs would be displayed. Then we went into a dive from over 20,000 feet, straight toward earth — and I was very much aware of the tightening G-suit pressure around my body. I was also cognizant of the fact that the earth seemed to be rushing up to meet us.

On the particular screen where the G-force indication appeared, I noted that it read 2.3 Gs. Then it was 2.7, 2.9, and so on. The green digits were steadily increasing. Then it was 3.1, 4, 5. My sight seemed to be impaired. My peripheral vision went grey as the blood rushed from my head toward my lower body. I could see the number 5.3, but it was as if I were looking at it at the end of a tiny tunnel. At that point I spoke on the radio to the major flying the plane and told him

I had had enough. My G suit was tight around me, and I understood implicitly that without it I would have been unconscious. As soon as we levelled out, the pressure dropped.

The pilot asked if I wanted to do an afterburner takeoff. I told him I did.

We circled over the runway, then dropped down and skimmed its surface. Twice. When I heard the pilot asking me if I wanted to continue, I no sooner agreed than we touched the tarmac a third time, and with sheets of flaming exhaust streaming behind us, the F-18 leapt into the air, straight up, in a mind-boggling, numbing ascent to 23,000 feet, more than four miles, all in less than sixty seconds. The acceleration pasted me to my seat and my G suit was tight around my abdomen and legs. Had I not been wearing it, and had it not been working as designed, I know I would have blacked out for sure. So without a doubt, Wilbur Franks's invention fulfills a need.

And today, it is not just used in fighter aircraft.

Every astronaut and cosmonaut who has left our planet in a space ship has worn a G suit, and and survived because of it. This was true when the first man in space, Yuri Gagarin, flew; when the first American, Allan Shepard, roared off the launch pad at Cape Canaveral, Florida; and since then, when every one of the men and women soared to the International Space Station. No doubt Wilbur Franks would be proud of what he left us, even though most who fly today do not even know his name.

13

SANDFORD FLEMING
Time for All the World

Sir Sandford Fleming has been called "Canada's foremost railway surveyor and construction engineer of the nineteenth century."[1] He was also one of the most creative men this country has ever known. In fact, we all make use of one of his ideas every day. That is because this is the man who gave us the concept of standard time, which is accepted now in every country of the world. And even though that invention alone makes us all indebted to him, there was a lot more to this amazing pioneer. He left his mark on the nation and on the world as few others have. And sadly, this being Canada, we never really granted him the praise he was due. Other countries put their heroes on pedestals; we forget ours. But every time we check the time, take a train, post a letter, or make a long-distance telephone call, we touch a Fleming connection.

Sandford Fleming was born in a small Scottish town called Kirkcaldy in 1827. He had little formal education — only six years, apparently — before he apprenticed with a local surveyor — a man who went from success to bankruptcy over a period of a few years. But because of this man, Fleming learned his trade, and in many ways became familiar with the ways of the world. When he saw that his town had an uncertain future because of troubling international events — the famine in Ireland and harvest failures and revolutions in Europe — he decided to go elsewhere. Shortly after he turned eighteen, he and an older brother left home, took passage on a sailing ship, and came to Canada. Because of horrendous storms at sea, they were lucky to survive the voyage across the Atlantic. Nevertheless, after a total of forty-four stomach-churning days on the ocean, landed on our shores in 1845, and ultimately made their way to Peterborough, Ontario. He then called this backwoods village his home for several years. He worked hard, took any job he could find, locally or in neighbouring towns, and saved what little money he could earn. The totals added up, however, and within three years he was able to help bring his parents to the area. They settled on a small, partly cleared piece of land outside of Peterborough, and, despite periods of wistfulness, never really regretted leaving their homeland. Instead, they, like their son Sandford, set about improving their living conditions and adapting to their new surroundings. As they did so, Sandford was also advancing himself in every way he could.

In many ways, the young man was ideally suited to this new land. He was strong, quick-witted, persistent, and daring. He was also a keen observer and a good judge of character. Perhaps that was why, soon after he met a delightful young woman named Ann Hall, he knew she should become his life partner. Subsequently, ten years after he first set foot on Canadian soil, he married her in a small church ceremony in Peterborough. Ultimately, the two would have nine children together: five sons and four daughters. The eldest was Frank, and we will hear more about his activities in subsequent years. Unfortunately, two of the children died young, and Ann herself passed away in 1888. Her husband outlived her by almost three decades.

One of Sandford's earliest surveying jobs in Canada had to do with mapping the harbour of Toronto. This summer and winter endeavour was successful and led to other related activities — particularly having to do with railways. It is said that there were only "25 kilometers of railway track in the whole country"[2] when Sandford and his brother arrived in Canada. That was where he next turned his attention, and work in railway-building would occupy him for years to come. And like his surveying of the Toronto harbour, the railroad work would be successful, but difficult and physically trying in ways that even he could never have imagined.

There was an extensive railway-building boom around this time and Fleming was determined to be part of it. His first projects involved railways in Ontario, and, subsequent to that, in Quebec and the Maritimes, as well. But, for Fleming, there was a much more exciting, and, he felt, a much more vital endeavour than either of these. Because this was at a time when the expansionist policies of the United States were paramount, Fleming knew that unless the land to the west of Ontario was more stoutly claimed as Canadian land, it would be threatened by American expansion. He had no wish to wake up one morning and learn that the Canadian West had become the American West. Already, some of the influential politicians and newspapers in the United States were advocating this. Fleming was determined to prevent it happening.

Sandford Fleming's dream (and the dream of others) was to build a railway from Toronto to the Pacific Ocean, so he threw all his vast energies into ensuring that this came about. By the time he was on course to do so, he was regarded as the best surveyor of railways in the country. He had influential friends in Ottawa and he urged them to push the idea of a railway west. By 1871, he accepted the position of chief engineer of the Canadian Pacific, a high-sounding title for Canada's most important construction project of the time — that of building a railway to the western sea. But, before any trains could run, hundreds and hundreds of miles of track had to be laid. And even prior to that, a proposed route needed to be surveyed, a task that Fleming accomplished.

In 1872, he, his oldest son Frank, and others set out to map the railway route. They spent months and months doing so, tramping over shale and muskeg, sleeping in tents and sleeping bags, under trees and rocky outcrops, and sometimes in the open where they were exposed to the wind, rain, and often snow. The work was never easy, rarely straightforward, and always fraught with danger from everything from bees to bears. Over time, all of the men fell, cut themselves, got sunburned, and sprained ankles, backs and shoulders. Yet they carried on relentlessly and often heroically. There were disputes over direction, terrain, and feasibility. Yet the survey was done and in time steel rails girded the country.

Then, on a bleak November day in 1885, the ceremonial last spike for the railway was driven home by a man named Donald A. Smith at a place called Craigellachie, British Columbia. In a famous photograph of the event, Sandford Fleming, sporting a bushy white beard and top hat, stands just behind Smith's right shoulder. Finally, the ceremony over, and his greatest railway project behind him, Fleming rejoiced with the amazing knowledge that now he could catch a train in Halifax and get off at Vancouver.

But before all this had become a reality, Fleming was occupied with other things, as well. As a prominent surveyor, he was well known in Ottawa. That was why the postmaster general of the day approached him to talk about, of all things, postage stamps. As surprising as it may be to us today, at one time Canada did not have its own stamps. That situation would come to an end however, when Fleming was asked to design the country's first stamp. The surveyor/railway man agreed to give it a try. It would be the first ever for the new country.

Fleming did drawing after drawing, rejecting idea after idea. And because he wanted a significant symbol of Canada on the stamp, he gave a lot of thought to what that might be. He rejected flowers, birds, farm animals, scenery, and polar bears. And even though Canada had eagles that were both striking and beautiful, the idea of depicting them was judged to be too American. He passed over flags, government buildings, and pictures of royalty for the first stamp, even though in the years to come lots of these would grace stamps drawn by others. He did do a drawing of Queen Victoria's husband, Prince Albert, but it was not the country's *first* stamp. He loved the land he now called home and he argued that its first stamp had to be unique. Finally, after rejecting suggestion after suggestion, he decided that the lowly beaver would be appropriate for the stamp — and ultimately, it was used. The little stamp would cost three cents. (Today, a so-called "Three-penny Beaver" in good condition is worth thousands of dollars to knowledgeable collectors.)

Finally, with the trains running and letters being mailed, Fleming moved often about central and eastern Canada and luxuriated in the sense of modernity his country now had. He always carried a pocket watch, and prided himself on being on time for any meeting devoted to one or more of his ongoing projects. He caught the train from Ottawa, heading for Toronto or Montreal often, and even though the times in each place might have been different, he was never late. This was because trains in those days ran on railway time, which was understood by Fleming and by most other travellers. The only problem with it was that each town along the route had its own time. If it was twelve noon in Toronto, it was a few minutes later in Belleville, and almost half an hour later in Montreal. As your train sped in either direction, you had to constantly change your watch to the new time. For example, should you have taken the train from Halifax, Nova Scotia, to Windsor, Ontario, you would have had to set your watch several times: "at Saint John, Quebec City, Montreal, Kingston, Belleville, Toronto, Hamilton, Brantford, London, and finally, Windsor."[3] In a way, every place along the line had its own time zone. But that was not as complicated as it was elsewhere. Around this same time, the state of Michigan, in the United States, had twenty-seven time zones.

But initially, Fleming was used to such complications and lived with them. There is a story about him however, and though it may be apocryphal, brings the point home. He spent a summer travelling by train in Ireland. Late one afternoon, he arrived at a small rural station called Bandoran, intending to catch the 5:35 train from there. When it did not arrive as expected, however, he checked a notice board and realized his train was scheduled for 5:35 in the *morning*. Accordingly, "Sandford Fleming, chief engineer of the Canadian Pacific Railway (CPR), would be a prisoner for the night at Bandoran station, and in the morning, miss his ongoing connections to … England."[4] It is said that while he waited, a plan began forming in his mind.

Fleming knew of course, that the earth rotates every day. He proposed dividing this rotation into twenty-four equal parts, with the first part running through Greenwich, England. There would be a one-hour time difference in each part. Initially, he met with plenty of objection. Many felt the old way was the best way, and that the way the railways operated was good enough. "Until the coming of the railroads, time had always been local time, the time by the sun."[5] When the sun was directly overhead, it was noon wherever you were. This idea caused lots of confusion, and Fleming felt that it impeded progress. He was sure that by dividing the world into twenty-four parts or zones, questions of time would be much less critical. That would eliminate the need for carrying more than one watch, or setting one's watch every few miles during travel.

A portrait of Sir Sanford Fleming painted in 1892. He was the chief engineer of the Canadian Pacific Railway, but is best known as the father of Standard Time.

But he continued having problems selling his idea and no government would take him up on it. He complained about this in many places, and ever wrote to the famous poet Rudyard Kipling about it. (Kipling was one of Fleming's many, many friends, some of them in quite influential places.) In his letter to Kipling, Fleming complained about governments in general, and about their not accepting things that were new — in any field. All governments refuse to consider something new, and convince themselves that, if it is new, "it is impractical, because it has never been done before."[6] The argument goes in circles, and is extremely frustrating. But Fleming had no intention of giving up.

He began compiling background materials and arguments for his position, radical as some said it was. He wrote paper after paper, extolling the idea, and finally, on February 9, 1879, at a place called The Canadian Institute in Toronto, Fleming read an important proposition extolling the virtues of Standard Time. No one now knows how many heard Fleming explaining his idea that day, but because he would not let his plan drop, it gradually became better known. In actual fact, he had intended to deliver the paper to the British Association for the Advancement of Science a year earlier, but the members of that august group refused to let him do so. After all, Britain had only one time zone, and thought Fleming's idea was useless because the people there did not use more than that. He was rebuffed with rudeness.

Again, however, Fleming did not give up. In fact, he more or less circumvented the Advancement of Science members by going behind their backs. At the time, the governor general of Canada was not only British; he was also Queen Victoria's son-in-law. And, pertinent to this case, he was Fleming's good friend. Fleming approached him on the matter.

The governor general listened with care and was intrigued by what Fleming proposed. He then decided to counter the short-sightedness of the Advancement of Science members by having copies made of the upstart Canadian's Standard Time proposal. Then he sent the paper to what he regarded were the most important countries of the world at the time, no matter where they were. There were several responses, but a particular one from Czar Nicholas II of Russia led to greater success.

The czar not only liked the Canadian's idea, he called for it to be explored, and suggested a meeting in Venice, Italy, to study time. This gathering came to pass, and Fleming finally got to deliver his paper. Subsequently, a much more important conference was held in Washington in October of 1884. There at long last, Fleming's idea won widespread approval, and Standard Time came into effect on January 1, 1885.

But this Scottish-Canadian jack-of-all trades always had within him more ideas than railroads, postage, stamps and time. Pierre Berton wrote years ago that if there could be a criticism of him, "it was that he had too many interests."[7] Nevertheless, there was another important role that the man played in his eight-eight years on earth. This time, it had not only to do with his adopted homeland; it was significant to Australian and New Zealand, as well.

Fleming had often travelled from sea to sea within this country and from Britain when he came here, and later, as well, so he knew how important communication was in linking various parts of the world. This fun-loving, gregarious character often looked at Australia and New Zealand and seems to have felt that those nations, no matter how advanced or worldly they were, were forever isolated below the equator in the far Pacific. He resolved to do something about that. After pestering Canadian and other governments for years about his idea, finally was able to see a communications cable laid across the Pacific, between Australia and New Zealand and Canada. The project took time, money, and was fraught with disaster often, but it did come to pass. The cable was finished and in use by 1902.

But, Fleming himself, this fascinating, cigar-smoking lover of good company and good wine finally was wearing down as an entrepreneur. He resigned from his railway duties, citing ill health, but never really withdrew from public life. He was knighted in 1897 while he was chancellor at Queen's University in Kingston. He did many things, contributed to his country and his world, and we should remember him. Sandford Fleming died in Halifax in 1915 and is buried in Ottawa.

14

SPAR AEROSPACE
An Arm from Earth to Space

Millions of people in countries around the world watched the television coverage of the American spaceship *Endeavour* being launched from Cape Canaveral, Florida, on the morning of Monday, May 16, 2011. This was the beginning of the final trip for that shuttle. It carried six crew members, one of whom was a young American-Canadian named Drew Feustel, embarking on his second flight beyond the bounds of our planet. Two years earlier, he had been one of the courageous astronauts who made repairs to Hubble; the big space telescope that, since its launch in April 1990, has been helping human beings see and learn about the universe that continues to expand before us.

On that mission, and again on this one, Feustel would be able to do his work because of a uniquely Canadian invention specifically built for the rocket he rode. That invention is something Americans call a Remote Manipulator System. It is a unique, unbelievably complex machine; the work of collective creativity and genius. It was conceived in Canada, built here, and now that the shuttle programs are over, will be placed in museums for all to see.

Canadians call it Canadarm.

Canadarms were standard equipment on each of the five shuttles that flew, and a more complex version was installed on the International Space Station, as well. That arm will continue to operate for as long as the station exists, although no one as yet is really sure how long that will be. Both the station and the shuttles that were used to build and service it have helped advance our knowledge of the universe. The early explorers found new lands; the space programs new worlds. Canadians have been partners in making this possible; working in many ways, over many years. Some of our contributions were small; others large. Some are widely known; others recognized by few. But of them all, the amazing Canadarm is in a class by itself. So is its story.

The first arm came about because the United States was building a new spaceship to replace earlier ones that had taken John Glenn around the Earth, and, in time, Neil Armstrong and others to the surface of the moon. Unlike those craft that were, in hindsight, functional, primitive, and quickly antiquated, the National Aeronautics and Space Administration (NASA), wanted a new vehicle for space exploration and, ultimately, something that would take human beings and equipment to an orbiting space station they hoped to build beyond the gravitational pull of our planet.

The old spaceships, or capsules, as they were often called, came back to earth under parachutes. Then they had to be picked up at sea, always a dangerous undertaking that held much potential for disaster. The fact that every one of those early astronauts survived pickup was due to luck as much as to the technology of the time. The only major close call happened to Gus Grissom, who almost drowned while *Liberty Bell 7*, the capsule he commanded, sank in the Atlantic Ocean. Unfortunately, he and two others would later lose their lives when a prototype of what would be the shuttle burned in a flash fire during testing at the cape. In retrospect, that accident illustrated that building a new spaceship and perfecting it would be fraught with problems. So would the creation of a lifting device for it; but that would be Canada's concern.

The purpose for building the space shuttle was because NASA wanted a reusable machine that would take off and land like an airplane. Having humans returning from outer space only to have to depend on a parachute for survival was no longer NASA's plan. But the months and years it took to create the shuttle were not easy ones. The project was extremely expensive, the work dangerous at times, and it involved thousands of people. As many great ideas were discarded as were used.

Primarily because of the expense entailed in the manufacture of the shuttle, the United States contracted out the construction of some of its parts. By spreading the costs to other stakeholders, the huge financial outlay was reduced for the host country. After much diplomatic consultation and intergovernmental co-operation, it was agreed that a huge lifting device for the spaceship would be supplied by Canada. The machine needed would be a kind of space crane. It would be vital for hoisting objects from what would be the cargo compartment, or belly of the spaceship. The crane itself would be anchored in the shuttle bay, and, when needed, could be pivoted out from there. Astronauts on the shuttle would be able to expend the "arm" of the lifting machine; remove cargo that was on board, or retrieve objects from outside the shuttle, and bring these back inside. Later on, such things as weather satellites were lifted from the shuttle, placed in orbit outside it, and conversely, hooked on to and guided back on board when their functional lifespan was over.

Planning for the first shuttle began long before it would be needed. However, following the initial landing on the moon, and the success of the subsequent ones, work on the shuttle was accelerated. So was the planning for the Canadarm. In 1974, Canada agreed in principle to build it, and then, a year later in July 1975, our government formally signed on for the project. Despite the fact that no design was available and no prototype existed, it is a tribute to this nation that the forward-looking scientists and engineers of the day were convinced they could construct a machine that would have to work perfectly in the harsh environment of space. They contracted to have the first Canadarm designed, built, and tested within five years. In this, they succeeded. By February 1981, their creation had been trucked to Cape Canaveral and was ready for installation on the space shuttle *Columbia*, which had not flown at all at the time. Its maiden voyage took place two months later, on April 12.

Columbia's first flight lasted fifty-four hours, circled the earth thirty-six times, and carried two crewmen. Landing was on a dry lake bed at Edwards Air Force Base in California. The spaceship lost sixteen heat-resistant body tiles during the mission, but the flight was still deemed acceptable. Seven months later, on November 12, it was ready to fly again, this time with the new Remote Manipulator System on board. In Canada, Canadarm's builders watched with a mixture of pride and apprehension. They need not have worried. The arm's flight was successful — as would be every journey any of the arms would ever make. That record is enviable.

Canadian Space Agency.

The John H. Chapman Space Centre, headquarters of the Canadian Space Agency near Montreal, Quebec.

But the months leading up to the inaugural flight were traumatic indeed. There were design problems at every juncture, but the overall idea for the machine had its basis in human physiology. In essence, the arm made for space shuttles is a reflection of the human one. There is a "shoulder," a "wrist," an "elbow," and a "hand." A tiny motor, the size of a coffee cup, drives each joint. However, unlike the human arm, this one is "15 metres long and capable of maneuvering payloads about the size of a bus."[1]

Because of the extreme temperatures in space, tiny electrical heaters are concealed inside the two long booms that are most visible to anyone looking at Canadarm. The arm has to move freely in different directions: up and down, sidewise, and in a circular motion. The "hand" of the device is the most different from the human one, but it still has to be able to clutch objects, such as communications satellites that have to be taken from or returned to the payload bay. The "hand" is really three snare wires that are capable of grappling an object. In order for this capture manoeuvre to work, each satellite to be retrieved had a kind of hook built onto its outer surface. Then, the "hand" of the Canadarm could grab the object in question and place it where needed.

The principal firm involved in the construction of the arm was a company known at the time as Spar Aerospace. There were subcontractors and an important impetus came from the National Research Council in Ottawa. Most of the "space crane" would be constructed in a facility in Toronto.

One of the main hurdles to be overcome lay in the fact that on earth, the arm was so light that it could not support its own weight. For that reason, for testing purposes, a large, extremely smooth floor had to be installed in the Toronto building. Air-bearing supports were then used to simulate various motions of the arm, but even so, engineers involved were never able to replicate exactly the movement expected in space. Nevertheless, they assured NASA that when finished, the arm would work according to the exacting specifications stipulated.

Testing, modifications, and more testing of the product took up countless hours by an army of specialists. There were advances, setbacks, disappointments, and headaches, but slowly, and with a kind of dogged determination, the Canadarm was built. From time to time during this period, teams of NASA scientists flew to Toronto and assessed progress. In doing so, they questioned the Canadian technicians for hours, determined to assure themselves that the tiniest directive was followed. The visitors were not disappointed.

Today, as we look at photographs of the Canadarm in space, we tend to focus on its "arm" section rather than on other parts such as the "shoulder" of the device. The long extension that we see is white and gold in colour and it often seems to glisten in the sunlight, particularly when viewed against the velvet blackness of space. This boom section is covered with many layers of insulating material, put there to protect it from the continuous fluctuation and extremes in temperature encountered as the shuttle circles the earth. Predominantly emblazoned on the white external coating is the wordmark "Canada." It was obviously put there to indicate the country of origin of the space crane – and most Canadians have a true sense of pride when they see it. The origin and location of the word is of interest.

Initially, little if any thought was given to the idea of our name on the arm, but when members of the design team saw European contractors putting identifiers on things *they* were building, it was decided Canada should do the same. The Canadarm people contacted their counterparts at NASA, and after a period of consultation, got permission to use our national wordmark on Canada's most important space invention. But the means to that end was neither easy nor immediate.

The wordmark had to be in a certain place. It must not distract astronauts who would be manipulating the arm. It had to meet myriad technical provisos. The material it was made of had to be approved by NASA. And of course, it had to fall within a specific size. After every implied and real objection had been overcome, the go-ahead came to affix the identifier. That was when personnel at NASA noticed that they had an unexpected problem with the whole matter.

No *American* symbol would be as readily visible, whereas "Canada" would be front and centre in every televised image of the arm in action. Because of that oversight, a painting of the American flag was hastily put in the shuttle's cargo bay. After all, taxpayers in the United States were footing most of the bill for this expensive and revolutionary space vehicle, and they had every right to fly their flag on it.

But the Canadarm got built and a few of the facts about it reflect just what had been achieved: "Weighing less than 480 kilograms, Canadarm can lift over 30,000 kilograms — up to 266,000 kg in the weightlessness of space ... using less electricity than a tea kettle."[2] In essence, the machine was a technical marvel.

From the time of its installation in the cargo bay of *Columbia*, until the final flight to the space station, the arm has been problem-free. It has been employed hundreds of times, often by Canadian astronauts. All of *them* praised the thing — but so did every other astronaut who ever used it.

Initially, a single arm was to be built. Then there were four others, excluding what was coined Canadarm2, the one that still operates on the space station. Two of the first five arms were lost when *Challenger* and later *Columbia* were destroyed. *Challenger* blew up seventy-three seconds into flight, as thousands of shocked spectators watched on the grounds of the Kennedy Space Center. Seventeen years later, *Columbia* disintegrated in the skies over southwestern United States. But, much more devastating, not just the shuttles and the two Canadarms were lost. So were fourteen young men and women who crewed those flights.

NASA photo.

Astronaut Chris Hadfield stands on the end of the Canadian-built Canadarm to work with Canadarm2 during its installation on the International Space Station in April 2001.

As indicated above, the shuttles were costly to construct. So were the arms. It is estimated that the outlay for the design, building, and testing of the first arm came to approximately $108 million. However, "the Government of Canada's original investment resulted in nearly $700 million in export sales."[3] But back when the first arm was being carefully anchored in the cargo bay of *Columbia*, neither the initial cost nor the long-range remuneration were the matters at hand. The focus was on the placement of the arm and its in-flight performance.

On its first flight, the two-man shuttle crew, Joe Engle and Richard Truly, noted what they thought might be some unexpected vibration emanating from the arm. This was traced to the spaceship launch, when the giant engines of the shuttle and the solid rocket boosters hurl the craft to the heavens. In the flash and roar that follows ignition, the reverberation is still enormous, and I can personally attest to this. I recall standing on the grounds of the media compound at Kennedy when the shuttle burst from the pad and the sound hit: "with a roar like thunder, and a jolt like a powerful punch to the chest, the massive wave of sound clobbers you, takes your breath away, weakens your knees, and shakes the soil on which you stand."[4] And this was *after* sound-suppression measures were taken to limit the impact. At the time of the initial launch of Canadarm, the shock was even more profound. For that reason, NASA quickly came up with a "water-deluge system, installed at the base of the launch pad that sprayed more than a million litres of water into the solid rocket boosters' exhaust at the moment of launch."[5] The idea seemed to help ease the launch vibration, and the shuttles and Canadarms both benefitted. There were some initial fears that if the vibration continued or became more severe, the arm would have to be jettisoned into space. Needless to say, that did not happen.

And so it was that Canadarm, our nation's foremost invention and contribution to the space program was a resounding success.

15

ARTHUR SICARD
Making Winter Roads Passable

In the late 1880s and after, a man named Maxime Sicard owned a dairy farm in Quebec, just outside the quiet village of Saint Leonard, on the island of Montreal. At the time, that settlement was tiny, with a population of less than three hundred people. It had one street, a few stores, and a large church that gave the place its name. The church was the focal point of the parish of Saint-Leonard-de-Port-Maurice, but because of its length, the full name was rarely used in everyday conversation.

Virtually all of the inhabitants of the village and the surrounding area were French-speaking. Most had lived there long before 1886, when the village was formally established and its first municipal officials elected. The first mayor was Louis Sicard, who was in office from June 12 of the founding year until January 21, 1901. It is not known if he and the dairy farmer were related, but in all likelihood, they were. They would certainly have known each other as the community remained about the same size for several decades. It was only in the 1950s that it really began to grow. Today it is an important and bustling part of the city of Montreal.

Just before Christmas in 1876, Maxime Sicard and his wife Justine had a son, whom they named Arthur. He grew up on the farm at Saint Leonard, but in later years would move into what was then Canada's largest city. By the time he took up residence in Montreal, however, he had invented a machine that he is still known for today. And as with many inventions, it came about because of a need. In this case, although the need was at the family dairy, it also existed throughout the area, elsewhere in Canada, and in far-flung locations around the world. Arthur Sicard lived to see the early dissemination of what he brought about, but he probably never could have envisioned how widespread and useful it remains to this day.

The need at the family dairy was felt from early December to late March, when winter storms blanketed Saint Leonard with snowdrifts that were as high as the fences at the farm. When young Arthur was a boy, there were no cars on the dirt roads, and in the winter these were often snow-packed, anyway, so getting around was never easy. People walked into town, rode on horseback, or travelled by cutter or sleigh, both of which were horse-drawn. Neither of these conveyances was particularly quick, however, nor were they always reliable. When snow was deep on the roads, walking was difficult, and horses sometimes floundered as they made their way. That was when running a dairy was particularly difficult. As we all know, milk does not stay fresh for long, and transporting it to urban customers was often a struggle. When the roads were blocked, deliveries could not be made, and the milk sometimes had to be dumped. Refrigeration as we know it today did not exist.

Young Arthur Sicard dealt with the problem every year, and even when the first snowplows came into use, they were both scarce and not particularly effective. But then one summer Sicard had what he thought might be an answer to the problem of winter roads. He noticed a piece of equipment that was being used to harvest a grain crop in a neighbour's field. Some accounts tell us that the machine in question was called a thresher, although this terminology may not be technically correct. What Sicard observed was likely a combine, although it served the same purpose as a thresher. The combine was invented in 1834 in the United States and was horse-drawn for years. Then, when tractors came on the scene, they pulled it. The first self-propelled combine was invented in 1911, and within a few years would have been in operation in the Saint Leonard area. This was the machine that intrigued the dairyman. He looked at it closely, and it gave him an idea.

Sicard noticed that the crop being harvested was fed into the combine as it moved down the field. A moving knife in front of the device did the cutting and then wide wooded arms that rotated parallel to the ground drew the grain stalks into the harvester itself. These stalks were shaken up until the precious grains went into a bin on board, while the unneeded stems of the plants were fanned away from the rear of the machine. In other words, something was drawn into the front and blown out the back. Some kind of similar device, Arthur Sicard thought, might also move snow.

In all likelihood, he was not aware of a kind of snow blower that had been built years earlier, prior to 1870. Robert Carr Harris of Dalhousie, New Brunswick, came up with such a machine, but his was for railways only, and it never gained widespread acceptance. The Carr Harris model "was mounted on the front of a locomotive and powered by the engine of the locomotive."[1] What Sicard envisioned was something that could be used on a road.

During the next few years, he thought carefully about what he had seen the grain combine do and he resolved to try to replicate the idea for snow removal. He built prototype devices, but none suited him. Sometimes what he welded together looked to be good indoors, but when he went to try it in snowdrifts outside, it would not start, plugged up, seized up, or proved to be totally inadequate. Each attempt was greeted with derision from friends who sometimes came to watch what was happening. Often they thought that the poor man was living in a dream world and his contrivance would never work.

But Sicard's efforts gradually paid off. By 1925, he had constructed a snow remover that did work. The thing would be called a snow blower because that's what it did: blades at the front of the thing drew snow in and then a large fan blew the snow out. In essence, every snow blower since, big or small, has operated on the same principal. But because Sicard's invention was built to clear the road from the farm, his machine was the size of a truck. It was not made for domestic use.

For the Sicard product, a truck body had a blower attached to the front and the fan that blew the snow got its power from an auxiliary engine mounted on the back of the chassis, to the rear of the vehicle cab. The truck was a four-wheeled drive affair, meant to have the necessary traction no matter how deep the snow to be cleared might be. When in full operation the machine could throw snow up to ninety feet in any direction.

Once he got the blower working to his satisfaction, Sicard manoeuvred up and down the farm laneway, clearing it quickly and easily. Then he turned onto the road in front of his property and succeeded in making the snow-packed right-of-way passable in no time. While this demonstration was underway, we can presume other farmers, including the former skeptics, came out to see the show. We have no idea what was said at the time, but we do know the results of Sicard's perseverance: the nearby municipality of Outremont asked him to use his snow blower and clear the streets of that town. The historical record tells us that this came about in 1927 — two years after Sicard began to build his machine.

He called his invention the "Sicard Snow Remover Snowblower," and either did not notice nor care that the name was somewhat redundant in nature. In no time, his snow blower caught on. Other municipalities in the Montreal area soon wanted their own machine. More were built, modified at times, then put to work clearing streets, parking lots, and other areas that had to be snow-free if possible. Word of the marvellous new invention spread. Crowds flocked to see it in operation, newspaper and magazine stories were written about it, and soon there was a backlog of orders from places far from Montreal. Inside of a year or so, this amazing method of moving snow was known across the country.

A local company manufactured the machine and from the start could barely keep up with demand. More workers were hired, and at a time when unemployment was high; building snow blowers was a godsend for many. Some who had been out of work for long periods of time got jobs and honed their skills constructing Sicard's invention. He formed his own company, gave it his name, and it still operates today, in Quebec and elsewhere.

Every fall, where substantial snowfall is expected, municipal maintenance staffs spend time getting their winter road-clearing equipment ready before the first storms actually arrive. The preparation generally includes snow blowers, some of which are understandably quite large. Today, some even operate on tracks in order to give them additional power in heavy snow. Then, instead of depositing snow to either side, they blow it into large dump trucks that haul it away. Depending on the location, by morning most of the major thoroughfares are clear, even after all-night blizzards. The feeling is that commuters have enough to contend with on the way to work; snow-clogged streets should not add to the problem. Surely, if Arthur Sicard was around to see such operations, he would undoubtedly be amazed.

But years ago, as word of Sicard's snow blowers spread and more and more of the machines made their appearance, the idea of building smaller units was broached. However, it was not until the 1950s that "walk behind" home units came into general use. Some of them were reasonably well built; others were just junk and owners swore off using them. However, the better ones got better, and they never went away. Now a relatively small number of companies make the machines, but with mass production, they are used in several places where heavy snow is a problem. For example, the Sicard company today exports to more than a dozen countries, with more signing on every year.

Today, there are snow blowers that are used for all kinds of tasks, ranging from the clearing of parking lots, to the clearing of airport run-ways, to the removal of snow from railway tracks. Most of the machines burn gasoline or diesel fuel; others are powered by electricity. Some are so small that they can be folded and hung on a garage wall for summer storage. Others are so big that they are hard to fathom. One kind in Nor-way can move twelve thousand tonnes of snow per hour. *That* would surely impress Arthur Sicard.

Undoubtedly, during the time he was experimenting and trying to construct his first snow blower, he must have skinned his knuckles, cut his hands, or perhaps even broke a finger or two in the pursuit of his goal. It is not surprising today, therefore, that snow-blower accidents have become much too widespread. Sicard may well have anticipated this, but there is no record of his mentioning it.

Nevertheless, accidents happen every year. Most are relatively minor; others much more serious; some fatal. The most common per-haps are strained backs, pulled muscles, or falls while clearing slippery sidewalks. The ones requiring the most medical attention are those involving hand and figure injuries. These incidents generally occur when the blower becomes clogged and the person operating it attempts to clear the machine. Such injuries only take a split second to happen, but the consequences can be traumatic. In fact, many result in finger, or even hand amputations. In a typical season, the United States Consumer Products Safety Commission estimates that there are close to six thousand snow-blower-related injuries that require medical attention each year. Since 1992 in the United States, nineteen people have died from snow-blower accidents. There are no figures currently available that involved Canadians, although some probably occurred. One accident that received a great deal of publicity happened in the United States, but to a transplanted and famous Canadian.

The captain of the Colorado Avalanche National Hockey League team, Joe Sakic, was using a small snow blower at his home during the afternoon of Tuesday, December 9, 2008. At some point, the machine became plugged with snow and he attempted to clear it. Suddenly, the auger of the machine came free while Sakic's left hand was still deep inside the blower. When the auger snapped forward, it caught the athlete's hand, broke three of his fingers, and tore tendons in his wrist. Later that evening, he had to have surgery at a Denver hospital to repair the damage. Fortunately, the operation went well, but the hockey player was lost to his team.

The morning after the incident, Jean Martineau, the vice-president of the hockey club addressed a hastily assembled group of reporters and said his star forward was "very, very mad at himself." Then the team boss added, "'I've never seen him like that, and I've known him for twenty-one years.'" As the news conference concluded, Martineau asserted the obvious about Sakic: "He knows he made a mistake."[2] The hockey world lost one of its finest players when Mr. Sakic retired the following summer. He then accepted a front office position with the Avalanche organization.

Another modern connection to the snow blower that Arthur Sicard probably did not envision lies in the fact that it has become a lucrative source of income for many — and not just in the manufacturing process. Today, no matter where snow blowers are found, they are looked upon as necessary by many home owners, businesses, and municipalities. Streets need cleaning. So do parking lots, driveways, and sidewalks, wherever snow is prevalent. Somebody has to do these jobs, and snow blower operators supplement their income by providing such a service. If the winter snows are substantial, the amount earned is commensurate with the number of times they have to go out. Probably if Arthur Sicard was with us today, he would be clearing busy streets himself. He taught us one way of coping with a yearly problem.

Not so long ago, a brief profile of the man was done by a Calgary radio station. The account covered all the basic information about Sicard and called him "an authentic snowman."[3] The description was creatively appropriate.

And there have been other kinds of recognition. A town south of the St. Lawrence River, near Trois-Rivières did not forget Arthur Sicard. In her book, *Canada Invents*, author Susan Hughes noted that "a street in Becancour, Quebec [is] named after him."[4] This surely is fitting, because that street has to be cleared every winter, and what better way to do so would be to use the invention that has made winter more tolerable in lots of places.

16

HARRY STEVINSON
Finding Plane Crashes and Saving Lives

Late on Sunday, August 26, 1951, Toronto Maple Leafs hockey star Bill Barilko and his dentist friend, Doctor Henry Hudson, disappeared as they were returning to Timmins, Ontario, from a weekend fishing trip to the shores of James Bay. The two were in a Fairchild 24, a small float plane owned and flown by Doctor Hudson. In the pontoons of the aircraft that day were the fish the two had caught during their brief getaway. Just before departure from a place called Rupert House, they fuelled the plane, bade goodbye to a handful of locals, and shortly thereafter lifted over the tree-lined shore and were gone. Sometime that evening, the plane crashed and the two were never seen alive again.

The largest air and ground search in Canadian history unfolded in the days and weeks that followed. This involved a vast array of searchers, twenty-eight military planes, and even more civilian ones. Some 325,679 square miles of rugged bush land were scoured in an attempt to locate the missing men, but to no avail. The operation was a frantic race against time, as sleep-deprived spotters scanned trackless terrain for any sign of the Fairchild. The tragedy was compounded when two search planes crashed, and while no one on either of these was killed, there were injuries. During the entire search, the news was never good, and one month less a day after it began, the military part of the operation was over, but Barilko and Hudson were still lost.

Then, eleven years and thousands of rumours later, an Ontario Lands and Forests helicopter pilot named Gary Fields, on a timber survey over the northern forest, spotted something shining in the bush, some sixty miles north west of Cochrane, Ontario. Another search was begun, this one lasting seven days, but it was a success, from one point of view: the crash was located. At the same time, however, the conclusion brought renewed heartbreak to the families of the lost men. The skeletons of Barilko and Hudson were found, still strapped in the wreckage of the plane that took them to their deaths. Later on, it was believed inclement weather was a major reason why the Fairchild went down.

Still later, I wrote a book about Barilko[1] and in so doing was in contact with many of the principals involved in this story. Two of them, Barilko's sister Anne, and Lands and Forests pilot Fields, had divergent reactions to the crash. Anne was devastated by the loss of her brother, and that was totally understandable. Fields was upset, but for an entirely different reason. Because even though he had seen something shining in the bush, he needed seven days to relocate the object. Both of these people told me that if there had been some kind of homing device on the crashed plane, it might have been found sooner.

Realistically, even if the downed aircraft had been located right away, perhaps both men might have been already dead. However, at least their whereabouts would have been known quickly and the agony of the loss would have been alleviated somewhat. Anne Barilko's eleven years of not knowing would have been erased. And, as for Fields, he endured seven days of intense personal and media pressure because he couldn't find the plane in the bush. Everyone he encountered in that period thought that he should have been able to go to the location of the sighting right away. Not so. The bush in the area was just too dense, too unforgiving, and too dangerous. There was muskeg, impenetrable underbrush, bees, blackflies, and bears. When the wreckage was located, a police helicopter had to land a mile from the site, and the investigating officers and others had to hack their way through the bush in order to get to the location to retrieve the bodies.

The Barilko-Hudson crash was a major story when it happened and few Canadians were unaware of it. Just a few weeks before he died, Bill Barilko had scored the winning goal in a hockey game to win the Stanley

Cup for the Toronto Maple Leafs. That feat, and a memorable photograph taken at the instant he scored, catapulted him into national prominence. And Doctor Hudson was a well-respected figure in Timmins and area, so when the plane carrying the two went down, the nation knew. And that certainly would have included a young electrical engineer working for the National Research Council in Ottawa.

Harry Stevinson was thirty-six years old then and he had been with the NRC since the end of the Second World War. His work in the aeronautical division had to do with solving problems relative to radios and other equipment. He also devoted much of his time to the testing of aircraft operation, specifically that of the glider. These two research interests were important to him and so was another aspect of his work: the installation and operation of radio beacons in planes. As early as 1939, he started envisioning a distress alarm in a plane, and, in case of a crash, the alarm would activate.

When the war came along, Stevinson joined the navy and his crash-beacon idea had to be put on hold. However, once he got himself established at NRC, he took his idea to his superiors. They were interested and encouraged him to pursue it. From then on, Harry Stevinson would be known for what he invented, more that for his other tasks in Ottawa. In a story about him in a University of Alberta engineering publication, he was lauded for "his brilliant inventions, many of them tested at the home laboratory in his basement, sometimes left his colleagues at NRC scratching their heads."[2] The article went on to describe the idea that was his most important: emergency apparatus to be installed in planes.

For Stevinson, the Barilko-Hudson accident, coupled with a later fighter-jet crash spurred him on to design the beacon he envisioned. And then, in an accident reminiscent to the two involved in the Barilko-Hudson search tragedy, a second plane looking for the lost jet also crashed in bush land. It had been flying low in order to try to locate the original aircraft. Neither jet was equipped with a crash beacon and that had bothered Stevinson greatly.

In fact, the idea of being seriously injured or killed in an air crash had been with him for a long time. He thought of such a scenario as far back as his teenaged years, when he went for a plane ride near his Bashaw, Alberta, home. As he and the pilot flew that day over some pretty remote terrain, he found himself wondering what would happen if the engine quit, if they ran out of gas, or if a freak weather occurrence forced them down. He trusted the man at the controls, or he would not have flown with him, but engine failure at two thousand feet was no laughing matter. Without power, they were going down. And even if they survived a crash landing, how would they get the word out about their situation, or their location? They could be on the ground in a minute or so, and then what? The whole aspect of being lost was bad; having little hope of being found was worse.

Stevinson "mused that a bird sitting on the plane would fly away at the earliest inclination of trouble and would survive the crash."[3] But neither he nor the pilot could do as much. He never forgot that thought and it came back to him often in the period after his bosses permitted him to try and build the plane beacon he imagined. However, having the idea in his head was one thing. Actually inventing the device was another. And because he was familiar with several other rescue devices that existed, he felt that "these were insufficient. If the crash occurred over water, the beacon would sink with the aircraft even if the crew escaped and were on the surface. Over land, the aircraft itself would block the signal if the beacon ended up buried under the fuselage, and the crash and any post-crash fires had the possibility of destroying it."[4]

So no, Stevinson knew that he had to come up with something better — something as good as the revolutionary car he built by himself when he was a teenager. Back then, his friends said that it could not be done, but he proved them wrong. The car in question was an extensively modified Model T Ford that looked more like a rocket ship than an automobile. It had seven forward gears, a long sleek outer body, and would go seventy miles an hour. When he toured around Alberta in the contraption, there were only ten miles of paved highway there. Nevertheless, because he could roar down the dirt roads with abandon, his friends became believers. Now, he intended to build a rescue beacon for planes and he convinced himself that the device would work.

Stevinson began his project by looking at every kind of emergency device that then existed. He found that some of them had aspects that might be considered and were perhaps worth keeping. Others were essentially junk devices that had been sold to private pilots to make them think they were safer if they crashed. Such things were about as life-saving as a lucky charm in the pilot's pocket, so the only ones who benefitted were the sellers of the things. All had to be looked at, however.

The next step was to make a kind of checklist of everything that would have to be included in an emergency beacon in order to make it worthwhile; to ensure that it was not just something that would fail with the first crash. The requirements were extensive and vital. The device "had to ride on the host aircraft in an inactive state for indefinite periods, but be ready to immediately start to work in the event of a crash; it had to survive any conceivable crash by falling clear as the aircraft crumpled; and it had to turn itself on automatically so that rescuers could find the crash quickly."[5] Stevinson believed that if he could come up with something that achieved all of these things, lives would be saved. Of course, whatever he invented had to be lightweight so that its presence on a plane would not affect the load factor in any way. And last, but perhaps as important as all of these factors, the device had to be cost-effective or it would never be used at all.

Stevinson decided early on that whatever he came up with should have no moving parts, as these would probably break on impact, or perhaps even before they fell from a plane. He knew he needed a transmitter, some kind of antenna, and then a casing or shell. He also decided that the device he came up with should be attached to the outer surface of the host plane and then, in time of trouble, break away by itself, fall far enough away from a crash, and survive. While he worked on the outer design for his creation, colleagues in another division of NRC built a small transmitter to be included in the package.

In his work, Stevinson was assisted by David Makow, a close associate. Together, they first constructed a flat, round, paper model of what would ultimately be called a Crash Position Indicator, or CPI. Together, they dropped this first device from a balcony in the building where they both worked and watched as the package tumbled over and over, seemed to slow somewhat, and then landed without damage. The next model was made of aluminum and it worked even better. In fact, "the model's performance, tested by releasing it from a speeding car, convinced Stevinson that the tumbling airfoil principle was almost ideally suited to this complex task."[6]

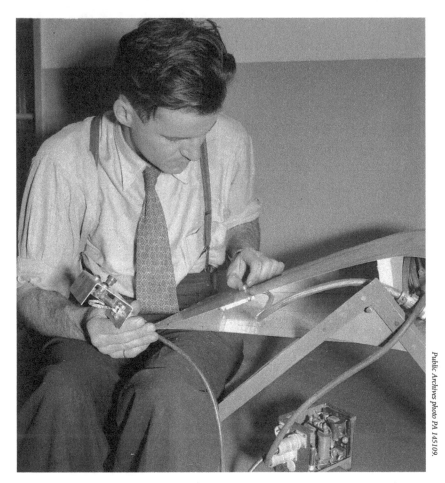

Public Archives photo PA 145109.

Harry Stevinson, inventor of the Crash Position Indicator. As a result of his resourcefulness, countless lives have been saved after air crashes.

They then constructed a fledgling CPI out of plastic and shot it from a cliff top on a rocket sled — into an abandoned gravel pit. The device survived. Finally, several drops were made from a variety of aircraft and they too were successful. From this point on, the positioning of the device on the outer surface of specific planes had to be considered. Stevinson and Makow thought of several locations, rejected some out of hand, but gave careful thought to others. And because all aircraft were not built the same, nor did they have the same shape, the place where they affixed the CPI necessarily varied a great deal. Even the speed of the plane had to be kept in mind.

It was one thing to build a small crash indicator for a small private plane that carried one or two people; it was quite another to construct something for a large, commercial, propeller-driven passenger aircraft. Obviously, none of these machines were the fastest in the world. However, putting a CPI on a jet, any kind of jet, was another matter entirely. Here, the great speed had to be kept in mind, and the particular placement of the indicator was doubly important. It goes without saying that a CPI on any plane had to be kept well away from the engines, but on a fighter jet the device had to be attached in such a way that it would remain in place when the aircraft flipped over, or performed high-speed tight turns in an aerial dogfight. All of these factors took weeks and weeks of study, but the resolve of the two men most involved never wavered. In time, the CPI being tested passed every obstacle around and Royal Canadian Air Force senior officers became interested. Initially however, the first use of the indicator was on a few planes operating in the far North, where rescue facilities were few and far between. "In one particular case, a plane crash in the Yukon mountains, location would have been impossible without the CPI. The internal antenna was able to get the signal between the mountain peaks to the search aircraft."[7]

During this time period, Harry Stevinson sought an American patent for his invention. His application, filed in Washington on March 26, 1957, was concise and clear, and the line drawings that were part of it were worthy aids to the understanding of the submission. The overall title of the document was straightforward: "Crash Position Indicator for Aircraft," and the purpose of such a device was cited at the outset: "The problem is to provide means by which an aircraft may be located as quickly as possible after it has crashed."[8] A few sentences into the submission, although not specifically named, are echoes of the Barilko-Hudson search:

> Such searches, often necessarily covering enormous areas of territory, may involve the employment of a large number of other aircraft for a period as long as a month. The expenditure of time and effort is very considerable and financial cost correspondingly high. Furthermore in some instances additional aircraft have been lost while searching.[9]

Then the application goes to the heart of the matter: "There is thus a very real need for a simple device that will indicate to searching aircraft the position of a crashed aircraft regardless of the condition of the aircraft."[10]

The document was taken under study by officials in the Washington office, and ultimately some months later, Patent Number 2959671 was granted to Stevinson on November 8, 1960. Reaction in Ottawa was one of muted satisfaction.

In due course, the RCAF decided they wanted the device for most of their planes, and particularly the larger transports that carried virtually everything the Air Force moved. Not surprisingly, Americans outside the Patent Office soon became aware of the CPI, and, after their own assessment, adopted its use. When searchers were able to locate a United States Air Force plane that had crashed at sea one night, they were doubly convinced about the wisdom of using this Canadian invention. Later on, Crash Position Indicators would also come into play in the Vietnam War. There, several lives were saved from as many aircraft, when the CPI led search personnel to crash sites. Often these were in snake-infested jungle locations and often in the presence of the enemy who shot at rescuers when they came on scene. Nevertheless, there was some success: the Americans were relieved and they sometimes wrote to Ottawa to express their gratitude.

But, in a way, the acceptance in the United States reached its pinnacle when *Air Force One*, the plane that carries the President had a CPI affixed to it. Leigh Instruments of Carleton Place, Ontario, who manufactured the device, were thrilled by such a placement and their sales success proved it. They probably could not have obtained a more noteworthy customer order. Soon, as well, European and other plane manufacturers adapted the CPI for both civilian and military aircraft.

But what became of the Crash Position Indicator in the years after its invention, initial adaptation, then widespread acceptance? In short, it is still around today, in a modified but greatly enhanced form. Now, it has been incorporated into the well-known device commonly called the "black box," which is found on all large planes the world over. (It is not yet mandatory for small aircraft.) The proper term for the instrument is Flight Data Recorder, but the purpose remains the same. It assists in the location of a crash, but now it also provides extensive information about everything from engine performance, to cockpit conversation to electronic instructions involving the operation of the entire aircraft. One of the first things done following a plane crash today is to locate the black box, which in reality is a bright orange colour so that it can be more easily seen by searchers in the detritus at a crash site. The black box today may not be as rudimentary as Harry Stevinson's invention, but neither are present-day communications, search, or investigative capabilities. Now we are able to have as much certainty about *why* a crash occurred as we have concerning its location.

For example, after Air France Flight 447, en route from Rio de Janeiro to Paris crashed on June 1, 2009, killing all 228 passengers and crew, it took almost two years to locate the black boxes. They were on the bottom of the Atlantic Ocean, hundreds of miles from land, but were intact and vital to the crash investigation. But, occasionally, black boxes themselves can cause controversy. That was no more apparent than in the World Trade Center disaster in New York City on September 11, 2001. Initially, the official 9/11 Commission Report claimed the black boxes were lost in the Twin Towers conflagration. Later on however, a New York City firefighter named Nicholas DeMasi wrote a book about how he and Ground Zero cleanup crew personnel located the boxes about a month after the disaster itself. That such a divergence of opinion could ever happen would likely never have occurred to Harry Stevinson when he simply wanted to be able to locate planes involved in air crashes.

But what of Mr. Stevinson himself? He left the National Research Council in 1979, and then worked for ten years as a consultant for what had become Leigh Industries, the original manufacturer of the Crash Position Indicator. He loved the work, but was never terribly concerned about himself. His "last five of those years were voluntary, because he could not be bothered to file timesheets,"[11] in order to get paid. Sadly, in his later years this brilliant, unconventional and modest inventor developed Alzheimer's disease and had to be admitted to a long-term care facility in Calgary. His contribution to aviation safety was profound.

17

JIM FLOYD
The Airplane That Should Have Been

One of the most promising Canadian inventions of all time ended terribly. The first jet transport plane built and flown in North America was destroyed before it ever reached its potential. Today only its nose cone remains, a neglected artifact in an Ottawa museum. The rest of this remarkable aircraft was torn apart and used to make pots and pans, and perhaps metal stepladders. The wheels ended up on a farm wagon and the hearts of the plane's designers, builders and flyers were broken and trampled. If ever there was a Canadian manufacturing tragedy, this was it.

The plane in question was called the Avro Jetliner, or C-102, and it was designed and built at Malton Ontario, where Toronto's Pearson International Airport is located today. First flown on Wednesday August 10, 1949, the machine was fast, comfortable, quiet, and far ahead of its time. In fact, it was already in the air and drawing rave reviews by the time companies in the United States got around to contemplating drawing board designs for such a craft. And while the C-102 was a marvel of engineering and human perseverance, it was also beautiful, and a marked contrast to anything that had come before it. And, importantly, it was not a weapons carrier.

Right after the Second World War ended in 1945, most manufacturers of military equipment turned their thoughts to peacetime needs. That was true across much of the globe, as well as in Canada. The big

Lancaster bombers that were vital during the conflict were no longer being made here and it was hoped that they and aircraft of their kind would never have to be used again. That was why, in the climate of the time, the idea of a jet airliner for civilian use came to the fore. A young man named Jim Floyd, already a noted aircraft designer in Britain, came to Canada to head a team to build the plane.

The choice of Floyd was an inspired one. He was brilliant, dedicated, decisive, and skilled. He was also a natural leader, respected by those he served and those he led. When he set up shop on the outskirts of Toronto in 1946, he and his associates would be less than four years away from seeing the plane they conceived leave the runway and fly. When that day came, it was the culmination of a drive to create that has rarely been equalled anywhere. Existing plant facilities were modified, workers were hired, and aircraft designs were drawn up, checked and changed, over and over again. Thousands and thousands of detailed sketches were done, then set aside and redrafted, only to be tossed away again when a better idea surfaced. But the project went ahead relentlessly, despite innumerable hitches, all in the face of a timeline that seemed to be cast in stone.

But why was this particular kind of plane being built?

The answer lies with the federal government in Ottawa and with what was then Trans-Canada Airlines, or TCA, as the corporation was generally called. In later years that company morphed into Air Canada, which as we all know is still our dominant airline. In 1946, TCA wanted a new plane — and not just any new plane. The company wanted an exceptional kind of aircraft that would carry both people and freight; sometimes together, sometimes not. TCA was owned by the government, so a fast, distinctive, and futuristic jet airliner would reflect well on the politicians who were keen to spend lots of tax dollars for it and on the company that would fly it.

The task of building the aircraft was monumental in those days before computers, when something called a slide rule was the most important tool in a designer's kit. No change was easily made, no structural alteration could be electronically evaluated, and no future fuel consumption rate could be readily assessed. And while fighter jets had been built and flown toward the end of the war, the connection between them and this large jet passenger plane were tenuous at best.

There were all kinds of designer specifications for the C-102, and most of them came down as directives from TCA. Initially, the airline brass wanted a thirty-six-seat aircraft that would fly at 425 miles per hour. Then they said the speed had to be five hundred. Eventually, passenger capacity was raised to sixty people, and cruise speed ended up being about 420. TCA dictated a range of twelve hundred miles; that the plane had to be able to land on four-thousand-foot runways; and that the engines must be of a certain type. There were all kinds of arguments over cost, load limits, stacking time over airports in case of landing delays, and increased range capability when diversion to a more distant location was needed because of weather or emergency.

Not surprisingly, in the face of a multitude of specifications that were often utterly tangential and often conflicting, Jim Floyd and his design crew became increasingly frustrated. However, to their credit, they carried on, and the aircraft took shape. Unfortunately, as the months went by, TCA officials never ceased to complain about something or other. They didn't like this or that. They told the builders to install things that had been faulted a day earlier. Ultimately, the company did not seem to care whether their new plane got built or not. In that light, it has been pointed out that they "lost interest before the first flight."[1] In other words, their support for the venture was largely superficial at almost every stage of development.

However, after overcoming every obstacle that had been put before them, the builders of the bird saw their creation fledged. On that memorable August morning in 1949, as the sultry heat of a southern Ontario summer hung over Toronto, the C-102 was eased from its hangar. In the left seat in the cockpit was Jimmy Orrell, a test pilot from England, seconded to A.V. Roe Canada, or Avro, in whose factory the plane was built. He would be the first to fly it. To his right, co-pilot Don Rogers would document every aspect of the memorable journey, and would also take the controls for a time before landing. Bill Baker, the flight engineer, also worked on the flight deck during the successful one-hour-and-five-minute flight. Then, shortly after a triumphant return to the airport, to the cheers of ecstatic workers, media representatives, and throngs of the curious, a reporter asked Orrell what it was like to fly the only plane of its kind in North America. "A piece of cake," he replied with a grin. His co-pilot later wrote of the C-102 itself: "The performance, handling characteristics and smoothness of flight were all outstanding."[2] It seemed that success was assured for this wonderful machine.

In the days that followed, there were lots of test flights, and while almost all were problem-free, one certainly was not. On the very second one, Jimmy Orrell went to lower the landing gear as he approached Toronto, but nothing happened. He repeated the manoeuvre — many times — but to no avail. He looked over at Rogers and the co-pilot tried to see if he could get the gear to budge, but he had no better luck. The main gear would not go down at all, yet the nose wheel of the plane did go down as it was designed to do. Then, however, that wheel would not retract. So there they were, with part of the gear up, the other down and locked. Urgently, the two men discussed their unexpected predicament and tried to figure out what to do next.

In short order, they were in contact with the tower, which quickly patched them through to engineers at the factory who were the most knowledgeable about the design and workings of the landing gear. These conversations continued for some time and one suggestion after another was made as to how to deal with the very serious problem aboard the C-102. As Orrell flew wide circles above the airport, additional attempts were made to both lower the main gear, or retract the nosewheel. Neither would budge.

The men on the aircraft flight deck and the engineers on the radio brainstormed about what to do next. Somebody even suggested making an emergency landing in nearby Lake Ontario, but that idea was rejected out of hand by the pilots. Finally, Orrell decided he would try a belly landing on the grass fringe of the main runway. At least, when they hit the ground, the grass would be a bit softer and more forgiving than the tarmac — or so he hoped. The idea was risky: the plane could break up, burst into flames and kill those on board — or the desperate manoeuvre might just work. Orrell figured he really had no choice. He circled the airport several more times to use up as much fuel as possible and lessen the chance of fire on impact with the ground.

Then, twice, he brought the big aircraft around as low and slow as he dared to just above the runway, looking for the best possible place to drop down. On each pass, Rogers and the flight engineer carefully checked the ground as they sped over it, looking for debris of any kind that might impede the landing and lead to disaster. Finally, at the end of the second pass, Orrell decided it was now or never. All on board hitched their seatbelts tighter as the pilot advised the tower of his intention. He approached the ground as carefully as possible, guarding against either stalling the plane or overshooting the place where he hoped to touch down. He, the others on board, and those watching below held their breath and waited as the plane came toward the airfield. Finally, Orrell checked that all was in order and put the jetliner down. A loud scraping shriek tore the air as the gleaming airplane made contact with the ground. But the nose wheel remained in place and the C-102 leapt forward, its belly and rear end dragging in the dirt. By this time, crash trucks were racing to keep pace.

They were never needed.

Orrell held to the controls with all his strength and to his immense relief the forward motion soon ceased. Once it did, those on board scrambled from the plane and sprinted away from it in case there was fire. Fortunately there was none, and apart from some shredded metal at the rear underside of the fuselage and scraping of the bottom of the engines, the damage to the C-102 was minimal. Repairs were made and it was back in the air a month later.

During the next several weeks, the plane was test-flown often, under varying circumstances and to a variety of destinations, such as Ottawa, Montreal, Trenton, and North Bay. Those who flew the aircraft and those who evaluated it never ceased to praise this flying wonder. It outperformed every yardstick it was measured against and impressed the most cynical of critics. The flight that gained the most publicity, however, came in the spring of 1950. That was because, on April 18 of that year, the aircraft was flown to New York City, where an enthusiastic and laudatory audience awaited it. Press representatives witnessed the arrival, and every reporter present extolled the machine — and faulted the United States for being so far behind in the production of such an aircraft. Canada, too, was praised, effusively and at length.

The *New York Times*, arguably the most important newspaper in America, covered the plane's arrival in a front page story on April 19. The *Times'* article told the readers that "the sleek new airliner received a prolonged welcome from the several hundred spectators who gathered at the airport to witness the first landing outside Canada."[3] New Yorkers marvelled that the flight from Toronto had taken only one hour and that the first-ever airmail carried in a jet plane had been on board. The receiving airport was Idlewild, now Kennedy International, and jet fuel had to be preordered there so that the tanks of the C-102 could be topped off for the return flight. This was because at the time, aircraft at any of the destinations the plane visited were still being fuelled with gasoline.

A bit later on, the jetliner returned to New York, at the end of a first-ever, direct flight from Chicago. This time, the airport used was La Guardia, in the borough of Queens. Rather amazingly, the fact that the plane would be coming there meant a great deal of local apprehension beforehand. No aircraft of its kind had ever landed there and the airport manager and his staff were afraid of what it might do. "They visualized the Jetliner as a fire-spitting monster which would melt the runways, scorch other aircraft on the tarmac and burn down the terminal building."[4] After some negotiation, they finally allowed the C-102 to land as long as the pilot promised to shut off the engines as soon as the plane was on the ground and agreed to be "towed into the terminal area if a conflagration seemed to be imminent."[5] The aircraft commander ignored all this nonsense and taxied to the terminal without incident. No one seems to remember what La Guardia officials had to say about the incident after it was over.

And the flight tests continued, to several destinations, and with unqualified success. As they did, American media outlets continued to praise the plane. For example, a newspaper in Rochester mentioned the remarkable Chicago–New York trip and claimed that the feat gave the United States "a good healthful kick in its placidity." The paper went on to praise Canadian enterprise and mocked American: "The fact that our good neighbor to the north has a mechanical product that licks anything of ours is just what the doctor ordered for our overdeveloped ego. Uncle Sam has no monopoly on genius: our products are not necessarily the best simply because we made them."[6]

Other largely similar comments were echoed in magazines, as well. A publication devoted to planes and flying declared that "most Americans believe that their nation has the greatest aviation industry in the world — an industry that embraces the most progressive manufacturers and the best in aeronautical brains. How then," the publication asked, "could first honours for a jet-powered transport go to the Canadians instead of our own fabulous aircraft industry? In the race to get a jet liner into the air, Canada won hands down."[7]

Even the eccentric aviation entrepreneur Howard Hughes, the president of Trans World Airways, loved the plane and wanted to build it himself or under license from Canada. He did get to examine the aircraft and even fly it several times, but his overtures relative to the development of the C-102 were ignored by the pertinent officials in the Canadian government. It was as if they had gotten their plane, saw that it was successful, but then couldn't figure out what to do with it. Instead, they decided not to build it at all, nor let anyone else do so. The project was suddenly cancelled, and the C-102 was dead.

When the news came down, there was widespread disbelief and utter consternation within the aviation industry and in the general public, as well. The shock to those who made the plane, those who admired it, and those who hoped to acquire it was total. Not surprisingly, every lame duck explanation the government came up with for their action was ridiculed, rejected, and openly contradicted.

But Ottawa was not swayed. Instead, those making the decision claimed that the jetliner was no longer a priority. They reminded the nation that a war had broken out in Korea and instead of a passenger plane, Canada now needed jet fighters, and quickly. For that reason, federal financing would now go to building fighter planes instead of commercial passenger craft.

The decision was bone-headed, arbitrary, and wrong, but it was the one given. In hindsight, the cancellation of the jetliner became "the major fiasco in the whole sweep of the history of Canadian technology." And further, as historian J.J. Brown has pointed out: "The decision to abandon the aircraft cost us billions of dollars in export earnings as well as incalculable world prestige."[8] Later on, another remarkable aircraft called the Avro Arrow was summarily cancelled, as well, and the loss was much the same. But that is another story. It would seem that at times we are a short-sighted nation indeed.

18

WILLIAM STEPHENSON
Pictures without Distance

Several years ago, a Canadian writer explained what a new innovation or discovery could mean for the person behind it: "Anyone who comes up with a badly needed invention — one that saves millions of dollars a year — gets rich, if he protects his idea at law."[1] That astute observation by author J.J. Brown probably could have applied to most inventors, but the specific reference was to a young man from the Canadian prairies. In this case, the inventor did get rich — very rich, before he was even thirty years old. In today's dollars, he would have been a millionaire many times over. In fact, for several years during the Second World War he even worked without pay because he did not need the income. The man's name was William Stephenson, and his hometown was Winnipeg, Manitoba.

Born William Stanger in 1897, he was just a child when his father died fighting in the Boer War. Because his widowed mother was suddenly left with four children and little income, she was unable to support them all on her own. For that reason, young William was raised by relatives and given their name: Stephenson. He retained it for the rest of his life. The boy was intelligent, motivated, and driven, and made his mark in many endeavours during a spectacular and successful career.

William Stephenson received his early education in his hometown at local elementary and secondary schools. Later on, he studied math and science at the University of Manitoba and subsequently taught there himself. Prior to doing the latter, however, he went away to war, and became a hero in doing so. This was the First World War, which broke out as he was finishing his final term in high school. Within weeks, he had traded the quiet of the classroom for the roar of the guns on the bloody battlefields of France. As one of thousands of foot soldiers in that terrible struggle to survive, he was gassed on two occasions; so seriously following the second that he had to be invalided back to England, his military career apparently over. But the young soldier was not ready to return to civilian life. He badgered everyone he thought might have some influence with the military, and in due course succeeded in getting into the Air Force, or the Royal Flying Corps, as it was called then. This in spite of the fact that he was seriously underweight, his face was pallid, and he looked ghastly. Nevertheless, he gave his all and learned to fly after only five hours of instruction. To the young pilot, the future looked more promising, and he soon became quite skilled in aerial combat above the rat-infested, filthy trenches he had so recently inhabited.

Bill Stephenson was a dependable pilot. He went on every mission he could, and came face to face with the enemy on many occasions. He shot down German planes, whether they appeared before him alone, or in company. Early on, after his own aircraft was hit, he crash-landed near his original takeoff point. Then, "he immediately got into another machine [plane] and insisted on returning to action. The next the squadron heard of him was that he had brought down two German fighters in flames. During the next two weeks, he had destroyed eighteen more."[2] Stephenson won several medals for such exploits, including the Military Cross and the Distinguished Flying Cross.

But then, trouble came to him.

On July 28, 1918, he went up by himself, only to see seven enemy planes attacking a little French reconnaissance aircraft. Stephenson took on the Germans, shot three of their planes down, but soon after he was hit, his plane crashed, and he was captured. Three months later, he escaped from the prisoner-of-war camp where he had been sent and made his way to England shortly before the war ended. Soon Stephenson was back in Winnipeg, a decorated flying officer, a veteran of ground and aerial warfare, and at only twenty-one, a young man who had already lived a lifetime of adventure, misadventure, and perseverance.

Intrepid Society.

The 1942 passport photo of Sir William Stephenson. A millionaire before he was thirty, he was a war hero who went on to invent the wirephoto.

But if Stephenson was bored by the sudden transformation to peacetime, he never showed it. Instead, he turned his mind to creative projects that would lead to recognition and financial success. The latter came quickly; the former took time. He began to examine radio broadcasting, then in its infancy, and beset with problems. There were lots of broadcasting companies, but the on-air product was still quite poor, largely because of static and overlapping frequencies that made relaxed listening impossible at times. Stephenson quickly noted that the primitive receivers, or radios, were far from satisfactory, as well. Later on, in an attempt to improve the product, he invested in radio manufacturing and did quite well in doing so.

However, one of the ideas Stephenson followed up on was to look into whether pictures could be sent, or transmitted, from one place to another in the same way as sound. Humans could talk on the telephone and the sound of their voices travelled on wires from one location to another. Similarly, a radio station could broadcast voices, music, and any other noise through the air, despite not needing wires to do so. Surely pictures could be sent, as well. Because he was convinced that there would be a market for photos transmitted in this way, he decided to see if he could figure out a way to make this happen.

He spent countless hours in his cluttered workshop, experimenting with one idea after another. Initially, nothing seemed to work as he hoped, but he never gave up. In his research, he learned of studies done many years before, in the 1880s, by a German scientist named Paul Gottlieb. Gottlieb had essentially the same long-range goal as Stephenson, and in his work, the German had invented something called a Nipkow Disc, a small, round, flat device with slits in it. Gottlieb used a single one of these and attempted to transmit pictures by using a strong light that would shine through the slits he had cut into the disc. However, the arrangement proved to be unsatisfactory. That was where Stephenson's inventive genius took over. Instead of one disc, he experimented with two, and found that doing so worked.

If he was going to send a picture, he "used a photo-electric cell and [the] two whirling discs to break down the image into uniform-sized dots of variable darkness, from very light to black, then into electrical impulses. Transmitted by telephone or radio, the impulses helped to create a crisp photo using a similar device at the receiver's end."[3]

By 1921, after he got his discovery working well, Stephenson began to make the rounds of Canadian newspapers, trying to interest them in his photo-transmission idea. At this time, several of the well-known papers that we have today were then in business, and while they published photographs of course, could not transmit pictures "over the air" to distant places. If a particular photo was going to be sent from Toronto to Vancouver, for instance, it had to be taken to a post office and mailed. Then, a few days later, after the mail arrived at its destination, someone collected the photo, and the receiving newspaper could publish it. All this involved time, effort, and even an element of chance because the mailed item was sometimes lost en route. Stephenson's idea not only eliminated the bother of mailing something, it made sending a photo almost instantaneous. The picture transmitted from Toronto could be published a few minutes later in Vancouver. A photo could be transmitted across the world just as easily.

Because Stephenson was so sure of the worth of his idea, he approached the major Canadian newspapers with it. However, much to his surprise — and chagrin — no Canadian paper was the least bit interested. They gave him and his invention little attention.

Historians have not told us what the inventor thought of such short-sightedness. We do know, however, that he quickly packed up in Winnipeg and sailed to England to see if he could sell his idea there. His first stop was at the *London Daily Mail*, and he did not have to go farther than that. All the senior editors he met were immediately excited about what they were being shown, as was Lord Northcliffe, the man who owned the paper. He quickly bought the invention from Stephenson, and published the first wire photo on December 22, 1922. For the young Canadian, the future was particularly bright, especially because even though he had sold to his first paper, he kept the rights to the idea himself. In no time other news organs bought in, and wire photo pictures became an important component at every large paper in Britain.

The possibilities of the invention were endless. A news photo taken in Melbourne, Australia, for instance, could be on the front page of a paper in England a few minutes later. The same picture might also be published in Montreal or Toronto because the big Canadian outlets soon realized that they had to have Stephenson's invention, even though they had unceremoniously dismissed it. There is no record extant however, of the investor reminding them of their earlier treatment of him. By this time, he was too busy selling his invention everywhere he could — and because it was so well-received, had little trouble doing so.

He also turned to other activities in the world of commerce.

In doing so, an observer pointed out, "he worked hard [and] made fortunes with apparently no trouble at all. He just accommodated every new idea, digested everything, and created out of what he absorbed."[4] He was an astute businessman and became involved in enterprises in Britain at first, and then in several countries on the European continent. In doing so, he travelled widely, made contacts easily, and observed the political situation everywhere he went. One of those places was Germany, which he noted with concern, was making clandestine moves to expand its military under a charismatic leader named Adolf Hitler.

From the outset, Stephenson distrusted the German politician, and watched with growing alarm the kind of militaristic movement the so-called Fuhrer was creating. And in contrast, Stephenson noted the appeasement moves in Britain, and the ever-more appalling concilia-tion efforts being made there by the largely ineffectual Prime Minister Neville Chamberlain. Fortunately however, the Canadian investor, by now a respected and successful entrepreneur, came to know Winston Churchill. For years, Churchill, although he was not in power, had been warning against the arming of and potential threat posed by Germany. In Stephenson, he encountered a like-minded man.

In due course, Churchill became prime minister in Britain, and from then on, his leadership and gritty determination proved to be what his country so badly needed. However, even after assuming power, Chur-chill neither abandoned his friends, nor ignored their advice. He had many dedicated individuals in his war cabinet, but he also went outside it for other purposes. One of those he turned to was William Stephenson.

At the time, the United States was not at war, and Churchill did everything he could to obtain America's help. While he conferred reg-ularly with President Franklin Roosevelt, he also felt the need to know what kind of sympathizers Germany had in the Americas. For that reason, he sent Bill Stephenson to New York to lead a clandestine intel-ligence network there in order to monitor Axis sympathies, not only in the United States itself, but in South America, as well. Stephenson was never technically a spy, but he assumed such a role, and performed admirably in obtaining information Churchill needed. Officially, the association headed by Stephenson was called the BSC, or British Secu-rity Control. It was "an organization whose power stretched beyond the United States into the whole of North and South America and its responsibilities covered far more than mere security matters.... Essentially, BSC was Britain's intelligence window in America."[5] Stephenson became a vital part of that structure.

His office was in Rockefeller Center, in the heart of New York City, but the contacts he developed and worked with were spread widely. In time, he became as knowledgeable as anyone about the activities of Nazi sympathizers in the United States and South America. His office kept tabs on hundreds of individuals, and, in time, intercepted letters, infiltrated pro-German groups, and built an intelligence network that was second to none.

During the course of his activities, Stephenson, along with one or two others were behind an Axis-monitoring and secret-agent training establishment in this country. The place chosen for the thing was east of Toronto, immediately south of the town of Whitby, on the shore of Lake Ontario. Here, land was acquired, buildings were erected, and a wartime school sprang into existence without any of its nearest neighbours knowing what was going on at all. Stephenson visited the place only once, apparently, having taken a night train from New York to Toronto, and then a car to the site.

The secret school was known as Camp X and even though it is long gone, it acquired a reputation that bordered on the fanciful. Many famous people are said to have gone there to be trained as spies, or secret operatives who parachuted by night into German-occupied locations in Western Europe during the Second World War. Ostensibly, movie stars of the time were among the trainees at the place, as was a writer who is well known today for one of his creations. The writer in question was Ian Fleming, the man behind the fictional agent James Bond.

It is said that Fleming spent time at Camp X learning all kinds of daring-do and then used what he absorbed to create the actions of the Bond character. But whether the writer was or was not at Camp X is probably immaterial. It is rather more significant that he likely used a real person as a model for Bond. That person was William Stephenson.

The inventor of the wire photo and the writer were not only known to each other; they became good friends. Ian Fleming admired Bill Stephenson and was somewhat in awe of his exploits. Because Fleming had his wartime training in secret operations, he likely looked at Stephenson as a man who was the best of the best in the field. No doubt, the 007 role is a dramatized one; anyone who has seen a Bond film knows that. And because Fleming looked at the Head of BSC as a self-made man of the world in a job involving secret activities, it was probably not surprising that the inventor from Winnipeg became James Bond.

And Stephenson would become someone else, as well. By the end of the Second World War, Winston Churchill felt indebted to the Head of BSC for all of his efforts leading to the defeat of Germany. For that reason, Churchill forwarded his name for inclusion on the British Commonwealth New Year's Honour List for 1945. Subsequently, William Stephenson was knighted by King George VI. A year later, when the United States bestowed on Sir William the U.S. Medal of Merit, he became the first non-American so honoured. Finally, on February 5, 1980, he was belatedly recognized by his home country when he was made a Companion of the Order of Canada. Sir William passed away in 1989, in Bermuda, at the ripe old age of ninety-three.

19

WALLACE TURNBULL
All for the Way of Flight

Some people probably called him a loner, but that would have been unfair. It is true; he worked alone, often for weeks at a time, but when he had the opportunity, he wrote for scientific journals, went sailing, read voraciously, and played tennis as well as many professionals. For all those reasons, Wallace Turnbull, a New Brunswick engineer and successful inventor, was unique. He is surely one of Canada's unsung heroes because his work in avionics changed the aero industry on a global scale. The man is dead now, but his contributions to aviation have outlived him and will continue to do so for the foreseeable future. He was a gifted and creative human being and his story is important.

Wallace Rupert Turnbull was born in Saint John, New Brunswick, on October 16, 1870. He died in his home province eighty-four years later. Because he got good grades in elementary and secondary school, and as he came from money, young Wallace had his pick of universities. He chose Cornell, at Ithaca, New York, and found that that institution suited him academically and socially. He did well there and in 1893 graduated with a degree in mechanical engineering. At this point, rather than returning to Canada or perhaps remaining in America, the twenty-three-year-old boarded a ship for Europe. He did post-graduate studies in physics at universities in Heidelberg and Berlin, Germany, and it was only then that he decided to go back to the United States.

Because his university credentials were excellent, Turnbull soon found employment with a company called Edison Lamp Works in New Jersey. This job lasted for six years and despite a praiseworthy work record, ended on an unfortunate note. Even though Wallace liked the job and the company liked him, his health was not good and he was forced to return to New Brunswick to recover. In the long term, the move was a good one. He bought a house in Rothesay, just outside Saint John, and gradually became well again. He also acquired additional land and a small barn adjacent to his home because he wanted to do experiments related to airplanes and flight, and he knew he would need the space. At the time, no plane had ever flown. The Wright brothers accomplished that historic first at Kitty Hawk, North Carolina, a year later in 1903.

Because she knew of her husband's interests, his wife seems to have feared that what he might do would annoy their neighbours, so she suggested he work alone in order not to cause unnecessary and embarrassing noise in the community. It was widely believed that airplanes would soon be flying, and Mary Turnbull knew there was much interest in making this happen. She did not want her husband to become what she called "a flying machine crank,"[1] like so many others who were building rudimentary planes and trying to get them into the air.

As soon as he got himself established in his workshop at Rothesay, Turnbull set about studying basic flight mechanics, with emphasis on the design of the wings that might be attached to a plane. He was particularly fascinated by the way air would pass over and under a wing, and how the wing structure could help cut through a headwind and assist in lift. In essence, he quickly realized that wing shape was critical for every plane that might ever fly.

Because he needed to study air flow, Turnbull realized that he had to devise some kind of apparatus to bring this about. That was why, after several attempts, he constructed what would be the first wind tunnel in North America. The structure was an ingenious and homemade effort. He hooked up a small, belt-driven, two-bladed fan to an electric motor. Each time the power was switched on, the fan pushed air "into a 16-inch square tunnel 6 feet long."[2] By using this crude device, Turnbull was able to assess how air currents passed over and under whatever it was that he placed at the end of the tunnel. He then went on to construct several small model wings and studied how the air moved when blown against them. Some of the models showed promise. Others indicated that they would hinder smooth airflow.

In all of his endeavours, Turnbull "was meticulous, thorough, and painstakingly committed to exactness and detail. He kept clear, concise notes on all his experimental work and used them to thoroughly work out the practical application of his ideas."[3] He even studied "the flight of birds, the conformation of their wings, and their air surfaces. This led him to test the aerodynamics of various forms of airfoils."[4]

This activity lasted for several years, but with the outbreak of the First World War, Turnbull felt he should somehow try to contribute to the Allied cause, even though his health was still not great. After discussing the matter at length with his family, he said goodbye to his wife and their four sons and went to England to help build the airplanes that were so desperately needed by the Allies in the skies over Europe. He was hired by a firm known as Frederick and Company and he remained with them until the end of hostilities. While Turnbull's duties encompassed several roles, his design and modification of airplane propellers became his most important contribution with the firm. The job was not without disappointment, however.

During his work term in the United Kingdom, he convinced the government of the day that what the English called an air screw, and we call a propeller should be modified so that greater loads could be carried on planes and they would become more functional. To that end, Turnbull invented something he called a variable pitch propeller. A major contract was signed to develop the thing, but before the Canadian profited in any way from his genius, the war ended and the politicians involved backed away from the agreement they had made. Fortunately, Turnbull still had one of the propellers in his possession. So, now that peace had returned to the world, he returned to Rothesay where he lived for the rest of his life.

Once settled back at home, he took the propeller he had invented and set about doing intensive studies using and improving it. He also constructed other propellers and studied them as they rotated in front of a fan. He changed the way they were built, altered the blade angles, the length, the diameter, the weight, and the materials used. During every experiment that followed, he made his usual detailed notes, measurements, and observations. But finally he had to figure out a way to move the propellers in such a way that he could learn how they would operate as if they were part of an airplane in flight.

To do this, he decided to build his own tiny railway!

It took quite some time to do so, but he carefully laid out tracks by a row of trees that edged his property. The trees helped eliminate motion caused by winds. The tracks themselves stretched some 350 feet and a carriage carrying a succession of propellers sped along them. The carriage, which was more like a trolley, could be made to move at a constant speed, and Turnbull would calculate propeller performance as it happened. As always, he kept careful notes of his observations. He ran the trolley down the tracks for days on end and studied the motion of several propellers. The better-performing ones were carefully modified, run again and again, and all modifications collated. No detail was overlooked and the "experiments allowed Turnbull to determine the effect of slip and forward speed on the efficiency of various types of propellers."[5] He later published his results to widespread acclaim within the aeronautical field. But now, he had to convince aircraft manufacturers and fliers that what he was proposing had merit.

Up until this time, all propellers in use remained static in flight. They could not be adjusted in any way, with the result that little cargo could be carried aloft. Because the power used to get a plane airborne was not the same as the power needed for straight-and-level flight once in the air, the aircraft of the time were extremely inefficient. As some have said, it was like only driving your car in low gear because there was no other gear. That was where Turnbull's propeller was a mark of genius. It *could* be adjusted in flight, so it was like second and subsequent gears in an automobile. And, luckily, the inventor still had his variable pitch creation with him and in time, he did get the attention of those who might try it out. Rather amazingly, Turnbull, despite his extensive interest in aircraft, was never a pilot. He could not simply put his new propeller on a plane and take it for a test flight himself. He had to rely on others to do that.

In 1923, following up on an invitation from interested individuals in the Canadian Air Force, Turnbull took his unique invention to an airbase at Camp Borden, Ontario, where it was hoped his plane modification could be tried. Accordingly, the new propeller was affixed to a suitable aircraft there and those involved prepared for a flight test. Unfortunately, it was not to be. Some ground trials were successfully completed first, but they were barely done when "a hangar at Camp Borden was destroyed by fire, and the only existing variable-pitch propeller in the world went up in flames."[6] A lesser individual than Wallace Turnbull would have been utterly discouraged by this turn of events, but he was unfazed.

He went back to Rothesay, to his workshop and little train. And because he had kept all the design specifics of the original propeller, painstakingly constructed another one that he felt was as good as or better than the one lost in the Borden fire. This time however, he built a backup — just in case. Then he tested both for weeks at a time. Sometimes the results suited him; sometimes they did not. He made more modifications and added a small electric motor in order to facilitate the necessary adjustments that would be needed to operate the propeller and modify the pitch for takeoff, for climb, and for straight-and-level flight.

The testing was extensive. It took almost two years to replicate his invention and bring it to the point where he was completely satisfied with it. During this period, Turnbull kept in touch with others who were also interested in flight advancement. Some of these individuals were of like mind; others were not. Alexander Graham Bell was totally supportive and the two became good friends over time. On more than one occasion, Turnbull visited Bell at the telephone inventor's home in Baddeck, Nova Scotia. Both men were interested in aerial dynamics, although they met challenges in different ways. In particular, Bell was more impulsive by nature than Turnbull, but the two benefitted greatly from each other's company. Undoubtedly, they left indelible marks on the Canada of their day, and on the country it would become.

Finally, in the spring of 1927, the time had come to test the variable pitch propeller in flight. Turnbull took the new one to Camp Borden and prepared for the critical unveiling of his invention. In bright sunshine, on Monday morning, June 6, the propeller was "fitted to an Avro biplane, piloted by Flight Lieutenant G.G. Brookes. The control of the pitch (blades) was by a small electric motor, mounted on the hub in front of the propeller." Then, to Turnbull's immense relief, "so encouraging were the air tests that a major development of the invention was launched without delay."[7]

Turnbull rejoiced following the positive flight tests, but soon had reason for disappointment, as is often the case where governments of any kind are involved. More follow-up was done using the same propeller and subsequently others with a slightly similar design. All worked and worked well, and Turnbull initially assumed that members of parliament in Ottawa would take note and help him promote such a unique facility. Air Force personnel were keen to help and they made sure the new propellers were quickly affixed to military aircraft. However, the politicians did not seem to be interested in even attempting to sell the idea outside this country. In essence, "the Canadian Government ... declined to have anything to do with its marketing."[8] Moreover, Turnbull himself was a modest man, never a braggart, and he lacked self-promotional skills. He would never push himself on anyone, or even take much credit for the products of his genius.

This was a man who had risen above three major disappointments. The first one went back to the failure of British politicians to follow through on their promise to develop the propeller when he worked in the United Kingdom. The second was the destruction by fire of the only prototype that existed at the time. Now, it seemed as if his own government had abandoned him just when success seemed assured. However, in each one of these instances, there is no record of Turnbull being depressed or even annoyed by the outcomes. He knew what he had done and he quickly continued to look for other areas where he might use his talents. Meanwhile, far-sighted individuals elsewhere saw the potential in Turnbull's propeller.

News of the revolutionary design was soon widespread in both the United States and England. Pilots and others in air industries in the two countries lauded the concept and clamoured for more information about it. Pilots from both nations came to Canada, were thrilled by what they saw, and what Turnbull's invention could do. They took the opportunity to fly aircraft with the variable pitch propeller while they were here, and in doing became quite enthusiastic about the modification. By the time these flyers returned home, a host of airplane manufacturing concerns were lining up to make the new propeller. All the while, Canadian government representatives stood on the sidelines.

When this hands-off nonchalance became quite obvious to Turnbull, he decided to follow up on business inquiries from outside the country. That was why he decided to enter into licensing agreements that did not involve Canada. Instead, he sold his patent rights to the propeller to the Curtiss Wright Corporation in the United States, and the Bristol Aeroplane Company in England. Both decisions proved to be financially beneficial for him, particularly the one with Curtiss Wright. That company sold propellers around the world and Turnbull was paid royalties for these sales. Yearly production levels grew, and then gradually leveled off until the Second World War came along. Then, many nations needed aircraft, and lots of them — in a hurry. This was no more true than in the United States where huge numbers of planes had to be built as part of the war effort. Suddenly, thousands and thousands of propellers were also needed, and Wallace Turnbull received royalties on all of them. By this time, however, he was deeply involved in other interests. For years, he had been fascinated by the hydroplane, the kind of boat that can skim across water at high speed. He also worked on bomb sights and torpedo screens during wartime. He remained active as he became quite elderly, when he delighted in being with his grandchildren, fixing their toys, and promoting and even playing tennis, the game he loved so much. At the time of his death, he was devoting his energies to studying how to make increased use of tidal power in the Bay of Fundy. In retrospect, though, the invention of the variable pitch propeller was undoubtedly his greatest contribution to aeronautical engineering and to the country he loved so much.

20

TOM PATTERSON
Saving a Town with Culture

The man was undoubtedly one of the greatest actors in the English-speaking world. For that reason alone, when Alec Guinness walked out on the stage on July 13, 1953, he gave instant credibility to the brand-new Shakespearean Festival in Stratford, Ontario. His first words: "Now is the winter of our discontent/ Made glorious summer by this sun of York" from *Richard III*, thrilled everyone before him. Not only were those two lines memorable, they were also the first spoken during a performance in the newest and most ambitious theatre in what was then the Dominion of Canada.

In front of Guinness, in the six-dollar seats nearest the stage, at either side, and in the rows at the back where tickets were a dollar, the audience that night sat transfixed as the lines of the opening monologue of the play hung in the air like a solemn harbinger of hope for the festival, the town of Stratford, and for the future of live theatre in the country. From that humble beginning, Stratford has become one of the largest and most important sites for the arts in the nation. Today, the festival has expanded exponentially. There are now five theatres. The performance season is long, the number of actors large, the array of ancillary personnel larger, and the impact on the local community and environs immeasurable. From the buskers in front of the Avon Theatre (and the author was there when Justin Bieber was one of them), to the bed-and-breakfast homes, the hotels, gas stations, restaurants, souvenir shops — indeed local and area businesses of every kind — owe much to the festival and the reality of what came to pass on that warm summer night some six decades ago.

But why Stratford and why Shakespeare?

To answer these questions, we have to go back, almost to the beginning of the tiny settlement on the banks of the southern Ontario stream called the Avon. And even then, the pretty little brook was not yet the Avon. The earliest settlers in the place called it the Little Thames, although that was changed in 1832 or thereabouts when the fledgling village got its name. At the time, Stratford was not much more than a clearing in the bush; a dot on the map within the million acres of land known as the Huron Tract. However, some twenty-five years after its founding, the town envisioned its future when two railways arrived. It was the rail connection that in time would lead directly to the fledgling festival that gave impetus to the institution we know today.

And even though the railways were important to the town, it also became known as an important manufacturing centre, particularly in the making of furniture. Most of the wood products were "made of oak and mahogany, but walnut was also used during periods when it was popular." The furniture produced "quickly established a Canada-wide reputation."[1] Ultimately, thirteen local plants produced more furniture than anywhere else in the country. And, as the workmanship was good, and because of the close proximity to the railways, the products were desirable and they could be shipped far and wide.

Unfortunately, with the coming of the Depression in 1929 and after, business fell off. That, and local hard times led to a furniture-worker strike in 1933. Some area food-processing employees were involved, as well. The demands of the two became more strident than companies would tolerate, and Canadian Army troops and tanks were brought to the town to deal with the matter. Fortunately, the upheaval soon ended, and "this was the last time that the military were requested during a strike in Canada."[2] Today, local people and those visiting Stratford can scarcely believe that such a thing could have happened in this quiet, pleasant, and picturesque little place. The Depression, the strike, and the rumours of another war all exacerbated the steady decline in the furniture business.

But the railway presence still remained strong and to most observers it looked as though it would last. By 1923, the two rail companies already there had merged into one, and ultimately, it too evolved into the Canadian National and then to Via Rail for passenger use as exists today. In 1871, "a locomotive repair shop came to town; it was expanded in 1889 and 1906."[3] From then on, the facility could do repairs to the largest steam engines then in existence, and, because of that, the workplace grew and grew, until eventually half the men in town worked there.

At least twice, the competing Canadian Pacific Railway attempted to drive one of their lines through Stratford. However, the move seemed unduly belligerent and because the tracks were to run "along the river through cherished parkland, the CPR was rejected in a public referendum and never came to Stratford."[4] Had the referendum not been held, an assortment of trains, along with the noise, dust, and interruption they would have caused would have been running through the very site where the festival is today. But ultimately, it was change related to the railway that almost brought about the death of the city it served. For this, Canadian Pacific could not be blamed.

The end of the steam era was a major cause, and, coinciding with it, the arrival of the first diesel engines. For Stratford and countless towns like it, the decline in railway use was also a factor. The age of the affordable auto had arrived. More and more people had cars and they drove to wherever they wanted to go instead of taking the train. Undoubtedly Stratford, with its reliance on the railway, was hurting. Workers were laid off and were told their plight was permanent. Many had to move and seek employment elsewhere. A sense of gloom settled over the town and no one seemed to know how to dispel it.

That was when a modern-day knight in shining armour rode to the rescue. His name was Tom Patterson and he was truly a visionary.

Patterson was born and raised in Stratford. He went overseas during the Second World War and upon his return, became a writer for *Maclean's*. As a journalist with connections beyond his hometown, he could not help but notice the rapid decline and decreasing workforce at the railway repair shops in Stratford. He realized, as did those who cared to look, that in time, train engine maintenance and overhaul would not be done there. But instead of lamenting the matter and doing nothing about it,

Patterson took it upon himself to act. With the clarity of hindsight and the perspective of decades in the future, one cannot help but marvel at the young man's astuteness and daring. He simply refused to accept the fact that a dream he had was only a dream. To him, it was a quest, a goal, and he was determined to see it acted upon.

From the time he was a teenager, Patterson thought his hometown should have a theatre where Shakespearean plays could be presented. And such a playhouse must not be a small, summer-stock venue where the presentations would be light, farcical, and forgotten. No, Patterson wanted a *real* theatre that would draw local, national, and international audiences. To him, it had to be a world-class establishment. So, with that end in mind, he set about trumpeting his dream. He gathered his friends around him, talked long and hard with them, and, inside of a few weeks, had them as enthusiastic as he was about his idea. But then the hard part began. He had to seek out and persuade local power figures, politicians, and opinion-makers.

At the time, he really did not know what he was doing, and later admitted this when he wrote: "Most theatres are started by actors or directors — in other words, by theatre people, whose total concern is for what will go on the stage. But because I did not know what was involved in producing a play, I was able to concentrate … on getting the people there to see what might happen on the stage."[5] It was that goal that drove him on.

Patterson's first contact was the mayor of Stratford, a man named David Simpson. Because the mayor was interested in promoting his town in every way he could, he rather quickly supported Patterson's idea. The theatrical dreamer then parlayed Simpson's name in meetings with others. The idea worked and gradually more and more people signed on; all believing the festival was a good idea. There was little money, of course, but at first that did not bother Patterson as much as it probably should have. He, the mayor, and then a festival organizing committee forged on. "By late 1951," Tom Patterson wrote later, "I *knew* that the Festival was going to happen, because I was going to *make* it happen."[6]

After getting $125 for expenses from the Stratford Town Council, Patterson went to New York in hopes of seeing the actor Sir Laurence Olivier and convincing him to come to Canada and be the first famous figure that would be part of the Stratford Festival plans. Not surprisingly, the idea did not work. And even though the young Canadian did chat up the idea with Olivier's secretary, he never saw the actor himself. He returned home empty-handed, with most of the council stipend gone. By this time, however, several Canadian newspapers were covering the story and generally in a positive way. The local paper, the *Beacon-Herald*, was among them, even though later on it was not always supportive.

At the time, Patterson worked in publishing in Toronto and because of that had been able to develop certain contacts. One of these was with a woman called Dora Mavor Moore, who was an accomplished director and actor. She liked the young man from Stratford and liked his festival idea, too. When he told her that Olivier was unavailable and would not be coming to Stratford, she suggested he contact Tyrone Guthrie. This man was perhaps the most pre-eminent Shakespeare director anywhere, but Patterson didn't know that "until he went to the library and checked *Who's Who in the Theatre*."[7] Dora Moore knew that if Patterson could convince Guthrie to come to Stratford that it would be a major cornerstone in the festival plan.

So Patterson sat down and wrote to the great man at his home in a village in Ireland. In his letter, he talked of Stratford, explained where it was, extolled its beauty, and asked Guthrie to come and advise the festival committee.

Then and since, there have been many who scoffed at Patterson's bravado, but it worked. "Guthrie agreed, on God knows what wild impulse, to come to this sleepy southwestern Ontario town,"[8] said one. But, despite the disbelief, the most important step in the process of founding the festival was in place.

Guthrie arrived alone and Patterson met his plane when it landed in Toronto. The two men introduced themselves, climbed into the Canadian's car and drove downtown. From the time the car doors closed, they became engrossed in an animated conversation about the idea of the festival. Patterson was thrilled by his guest's demeanour, and much later described his feelings at the time. "I was so very, very impressed with the great director! I knew by his probing questions that he was just as excited as I by the idea of the Festival and was obviously keen to get involved."[9]

From then on, it was Guthrie's festival. He wanted to see everything, meet everyone involved, look at the intended site, and go over what Patterson and the organizing committee intended. These activities took two weeks in Stratford and other locations in Ontario before Guthrie returned home to Ireland. Not long after that, Patterson flew to England and met with Sir Alec Guinness just prior to the actor going on stage in a play in the West End in London. Guinness, who had a family connection to Guthrie, thought the matter over for a few days, and then agreed to come to Stratford and be the headliner. As yet, no specific play had been chosen for him. Some days afterwards, an eminent designer named Tanya Moiseiwitsch along with Cecil Clarke, a stage manager, signed on, as well. By the time Patterson made his way back to Canada, the nucleus for the unprecedented theatrical adventure had been solidified.

Then Irene Worth, an American film star, was hired as the female lead. Shortly after her subsequent arrival in Stratford, she told Tom Patterson to ignore some of the finer points of her contract — such as the provision of a manicurist, a driver, and a full-time hairdresser. She merely asked if he could get her a bicycle. He did so, and she rode it to and from rehearsals and performances every day. In such ways, she charmed everyone who had any contact with her.

Not surprisingly, money was always a serious concern (even for the Worth bicycle), mainly because there was none. One of the major banks flatly refused to loan funds to the fledgling festival organizers, but, fortunately, another agreed to do so. Small grants came from other sources, including the J. Arthur Rank organization and the Ontario Department of Education.

In time, the physical aspect of the festival theatre began to fall into place. The stage was the first thing designed, followed by the rest of the building that would enclose it. While these elements were being detailed, Guthrie and Patterson, who by now were working full-time on the project, began considering and interviewing potential actors. Alec Guinness would be the main lead, of course, and the play *Richard III* was selected for performance, in addition to the lesser known *All's Well That Ends Well*. Programs at Stratford have expanded greatly since those first days and now include works by various authors on a wide range of topics. Initially, however, the festival was purely Shakespearean.

The quest for financial support was never-ending, but, through the generosity of Stratford citizens and others, the project continued. The local Gaffney Construction Company was the main builder and a man by the name of Oliver Gaffney pushed his men to what some called a super-human level. This was because, as with almost any major building project, a definitive deadline exists, and there is generally a final rush to make it. The huge tent that would be the roof of the theatre came from Chicago and prior to the structure's erection, play rehearsals were carried out at first in a building at the Stratford Fairgrounds, and then inside the tent once it was in place.

The arrival of the blue tent, one of the largest in the world at the time, was almost last-minute, and there was controversy over land ownership when it came to the exact placing of the poles that would hold the giant structure aloft. Yet somehow, despite myriad last-minute glitches, the canvas cover that would house the theatre was put in place. Teams of men on every side hauled on long ropes to raise the structure in a kind of pantomime that was as necessary as it was effective. Scores of the curious came to watch; spectators and reporters took pictures, and finally, the venue for the Stratford Shakespearean Festival was almost ready for use.

The date for the official opening had been publicized well in advance and tickets for the first performance were quickly sold out at the local store that handled them. In no time, tickets for *all* the performances that summer were hard to come by. Even some of the most influential newspaper theatre critics had to take inferior seats, but they did not seem to mind. When *Richard III* opened that first night, Alec Guinness was the actor everyone wanted to see and his performance was praised, as was the play itself. *All's Well That Ends Well* was also a hit. Years later, Herb Whittaker of the *Globe and Mail* would reflect on the opening season when he said, "The first season's *Richard III* provided the most exciting night in the history of Canadian theatre but the second night's *All's Well That Ends Well* topped it."[10]

The plays in Stratford were performed in the tent until the permanent theatre was built prior to the 1957 season. The time in the tent was generally memorable, although difficult in some instances. The heat of the Stratford summer made for some uncomfortable evenings. No air-conditioning as we have it today cooled the canvas structure, so both the performers and the audiences sweltered together. And occasionally, when the heat was not a factor, rain was.

Photo courtesy of the author.

The Stratford Festival has been an overwhelming success. In this picture, two men are memorialized as they pull ropes to lift the canopy of the original tent. In the background is the festival theatre.

From time to time, heavy downpours and the rumbling of thunder outside made the words from the stage hard to hear. But the rain did add an unexpected and rather amusing element. On occasion, small holes in the canvas let tiny droplets of water fall on the patrons below. And sometimes these were more than droplets. The author was there one night when this happened. I remember sitting in my seat, entranced by the play that was unfolding in front of me. As a heavy rainstorm swept Stratford, I noticed a couple of people to my right moving rather awkwardly to one side. Then a few others joined them. In the reflection of the stage lights, we could see a small but steady stream of water coming from the roof. A few seats were cleared as the play went on. I watched the rest of it from a step in an aisle, near the back. Unfortunately, I cannot recall the name of the performance, but I do remember the rain. The disruption it caused was rather fun.

Then, in preparation for the 1957 season, the permanent Stratford Festival Theatre was built. The tent had fulfilled its role; now the suc-

Photo courtesy of the author.

A life-sized statue of William Shakespeare sits outside the Stratford Festival Theatre. His play, *Richard III*, was the first ever performed here.

cess of the festival decreed that canvas was in the past. A new theatre was designed and it was built over the site of the first one. A superintendent with the construction crew that worked on the enterprise was a man named Jim Brown. He is elderly now, but his memory of the time is as vivid as yesterday. "When we moved in there, the tent was rolled up, over on one side," he says. "It weighed about five tons, but it was packed up and hauled away on a large truck. Then we started to build the new theatre above the stage and a few seats. It was a big job, and, as usual, there was a shortage of time. And there was not a square corner in the building."[11]

Brown is certainly correct in this. There is a specific reference to the shape of the structure in the archives of the festival. "Designed by architect Robert Fairfield, the new building would be one of the most distinctive in the world of the performing arts, its circular floor plan and 'pie-crust' roof paying striking tribute to the Festival's origins under canvas."[12]

But Jim Brown recalls other things as well.

The hammering and sawing went on as the building took shape. In the latter stages of construction, many finishing touches were underway while play rehearsals were in progress on the stage. "We must have been annoying to the actors at times," Brown admits. "But some of the people in those plays could certainly hold their own. I remember Frances Hyland, a beautiful young actress who was there at the time, putting a couple of our guys in their place. She was wonderful, and we soon learned that there was no nonsense when she was around."[13]

And the permanent festival theatre finally opened on July 1, 1957. The play that night was *Hamlet*, with a young, suave, talented, up-and-comer named Christopher Plummer in the lead role. And so to this day, Plummer, now an Academy Award-winning actor, has matured, as has the festival, to which he has brought much credit. I saw him that year when the theatre opened, and recently was enthralled by his solo performance in *A Word or Two*, and could not help but think back to the late Tom Patterson and the legacy he left. Christopher Plummer, arguably the finest actor in the world today, is just as good as Alec Guinness was in *his* starring role in the festival's first year — so long ago.

21

TIM HORTON
AND RON JOYCE
Double-Double Empire Men

It was shortly after 4:00 a.m. on Thursday, February 21, 1974, when the first Ontario Provincial Police radio report came in. Details were sketchy, but there was mention of a sports car that was really flying on the four-lane Queen Elizabeth Way (QEW) near Hamilton. A bit later, a second call came from farther along, at Vineland, from an officer who had unsuccessfully attempted to catch up to the speeding vehicle. He was sure the sports car was going at least 100 miles an hour. Soon afterwards another cruiser gave chase, but not for long. At 4:30, the speeder lost control and crashed his car at St. Catharines. The impact demolished the vehicle and killed the driver. His name was Tim Horton.

Because Horton was a Canadian hockey legend at the time, his death shocked the country when it occurred, and for a long time afterwards. Everyone wanted to know the details of how the accident happened, where it happened, but most of all — why?

The seeds of the tragedy were sown much earlier, but the more immediate sequence of events began the previous morning at a Buffalo Sabres hockey practice. The forty-four-year-old Horton was playing defence for that National Hockey League team, after spending most of his illustrious career with the Toronto Maple Leafs. The Sabre practice was uneventful until Horton went to block a slap shot from near the blue line. The puck deflected off a stick, hit him in the jaw, and broke it. Then, despite the fact that he was in severe pain in the hours that followed, Horton still played the first two periods of a game that night in Toronto against his old comrades. And even though the aching jaw kept him out of action during the third period, he was still named one of the three stars of the game. However, the aging veteran was greatly annoyed at himself because Buffalo lost and he felt he had let his team down by not playing the last period. After the game, he showered, got into his street clothes, left the rink, and climbed into his 1973 DeTomaso Pantera sports car. Punch Imlach, his coach for the Sabres, had given the car to Horton as an inducement to get him to play one last season for Buffalo. Imlach would later admonish himself for doing so, because in a way that car, one of the most distinctive and sought after on the road at the time, contributed to the athlete's death. He loved it; loved its handling, performance, and most of all, its speed.

Tim Horton loved to drive fast and this time was no exception. The night he died, he travelled west from Toronto to Oakville, and then stopped in at the head office of the donut-and-coffee company that bore his name. He was still there when his partner Ron Joyce showed up. While Horton held an ice pack to his throbbing jaw, he sipped vodka and discussed business affairs with Joyce. The two talked for much longer than they may have intended, but because Joyce had to be at an early meeting in Sarnia in the morning, he curtailed the session around 4:00 a.m. The two men then left Oakville — Joyce heading home to Hamilton, and Horton to Buffalo. The police reports of the speeding Pantera soon followed.

At the time he crashed, Tim Horton would have had to navigate a traffic circle on the QEW at the St. Catharines Lake Street exit. (The highway configuration has since been changed, and the dangerous traffic circle is no longer there.) Vehicles on the multi-lane QEW had to slow, follow a concrete wall around the circle itself, and then resume speed on the expressway. Undoubtedly, Horton was going too fast to make the move safely. One expert explained what happened that night. The speeding car "hit an elevated sewer grate and flipped several times, throwing Horton from the vehicle. His body was found almost 60 metres from the crumpled wreck of his beloved Pantera."[1] Later, highway-accident-reconstruction personnel noted that among other findings, he was not wearing a seat belt at the time of the crash — but in all likelihood, this would not have mattered, anyway. An autopsy on the body determined that Horton's blood alcohol count was twice the legal limit for driving in the province of Ontario. Inexplicably, the autopsy report was not made public until thirty-one years later, in 2005. Then, when it finally did come out, it stirred up all the same theories about what happened to this Canadian hero on the night of his death.

With the hockey player's passing, the Tim Hortons coffee-and-donut business was suddenly without one of its key figures and Ron Joyce had to bear the burden of continuing alone. It was not easy, but he did so admirably, and the success of the chain would reflect on his efforts at the time of Horton's death, before it, and since. But, we might ask, how did this iconic and innovative Canadian company come to be? The answer, in its simplest terms, was because a hockey player and a cop shared a dream and made it come true. Their stories are of interest, both in their sameness and in their diversity.

Tim Horton was born into a hardscrabble household in Cochrane, Ontario, on January 12, 1930. His father was a hard worker who struggled to put food on the table in the difficult years prior to the outbreak of the Second World War. That meant that he had to be away for long periods of time, doing a variety of jobs, anywhere he could find them. It was not until he obtained employment with a northern Ontario railway that there was relative family stability. There were two children, Tim and his brother, Gerry, and both loved sports. Tim was skating by the time he was four and playing pickup hockey with his brother and their friends at six. He concentrated on hockey during the winter and in summer "stayed in shape by lifting makeshift barbells made of pipe and old metal signs." Then, "by the age of twelve he was full-grown. Standing at 5ft. 9in, he did not grow to be a tall man, and was never a heavy man, but he was strong."[2]

Horton played minor hockey in northern Ontario until he was scouted by the Maple Leafs, and moved to Toronto to attend St. Michael's College and play with their Junior A team. He excelled there and in due course made the Leafs. He never looked back: He played more than a thousand games with Toronto, won four Stanley Cups there, made the All-Star teams several times, and was posthumously named a member of the Hockey Hall of Fame in 1977. In the course of his career, he made friends wherever he went, and no more so than among the men with whom he played. At the time of his death, former teammate Allan Stanley said of Horton: "He was the finest man I knew, on or off the ice. He was a great leader without a mean streak in him. Playing with him was a wonderful experience."[3]

Arnie Lee, Wikimedia Commons.

Tim Horton when he played defence for the Toronto Maple Leafs. Here, he is sitting in the penalty box at Madison Square Garden in New York, during an NHL game in 1965.

With such a personality, it is perhaps no great surprise that the fast-food business he entered would become a success. The first Tim Horton donut outlet opened in Hamilton in 1964.

Ron Joyce was born in Tatamagouche, Nova Scotia, in 1930, the same year as Tim Horton. And, like Horton, who was often without his father because the man was away working, Joyce too was without a father, but in a tragic sense. Willard Joyce was killed in an accident when Ron was three and his mother was left with the challenge of raising a daughter and two sons on her own. She did so in those early years by working wherever she could and subsisting on a small widow's pension. Young Ron was often sent to live with relatives for periods of time because of his mother's lack of money. In those days, the family lived in a small house with neither running water nor electricity.

When he reached his teenage years, Joyce dropped out of school and over time did farm, factory, and construction work in his home province. Then, one day he decided there had to be more to life, so he went to Hamilton, Ontario, on his own. He loved the city and was never unemployed there, but after two or three years got restless again. This time, he joined the Royal Canadian Navy and spent five years circling the globe and visiting places he never dreamed he would see. Then he returned to Hamilton and joined the police force. That job suited him and by all accounts he was good at it. He was also well-liked by his colleagues, and he had the reputation of dealing with those he encountered in a fair and even-handed manner. At the same time, however, he was never a pushover.

Ron Joyce worked as a Hamilton police officer for nine years, but as his salary was never more than $100 a week, he started looking for some way to supplement it. He ran a Dairy Queen for a while in his spare time, but one day noticed a former gas station at 65 Ottawa Street that became a little store called Tim Horton's Do-Nuts. This was the hockey player's first venture in Hamilton. A franchise in the place was for sale at the time, so Joyce bought it — even though others told him it was a bad investment and would never fly. He even admitted at the time of the purchase, that he "had never met the man after whom the store was named."[4] But as history would prove, it was this purchase in 1964 that was the real beginning of the national chain we know today.

The first Tim Horton's Do-Nuts opened at 65 Ottawa Street in Hamilton, Ontario, in 1974. It was a converted gas station and remained so until 1979, when it was torn down and replaced by a modern Tim Hortons.

The Hamilton store was now a donut outlet because Tim Horton had been looking for some way to ease into retirement after professional hockey. "He always worked between seasons at some job or other to help the family get ahead," his wife Lori recalled, "and for most of his playing days he was preparing, in one way or another, for life after hockey."[5] And while Horton's pay might have been adequate for the time, it still fell far below the salary of even journeymen National Hockey League players today. He knew he needed something to fall back on when he hung up his skates for good. It was for that reason that he had dabbled in an assortment of businesses, but none of them had flourished. Some of these were restaurants in places as far-flung as North Bay, Ontario, and Toronto. However, his wife recalled that he was always fond of donuts and had seen little outlets for such fast food in the United States. "For years and years," she said, "on our trips down to visit the family in Pittsburgh there was a little donut shop in Erie, Pennsylvania, which Tim liked so much he would go well out of his way just to stop there."[6] This routine made him start to wonder if a similar concept might work in Canada. That was why, when an opportunity to try such a venture materialized in Hamilton, he and a partner took it. Then they advertised their franchise and Ron Joyce bought it.

As Joyce would later admit, those first few weeks and months running the business were not easy ones — even though they were reasonably successful. There were problems with baking, equipment leasing, recipes, operating procedures, staff expectations, and even bakers who were poached by other restaurants. The problems came to a head when Tim Horton and the partner whom he had dealt with before Ron Joyce went their separate ways. Subsequent to that, Joyce and Horton were the two men responsible for the business.

By this time, as well, Joyce and Horton had become good friends.

As was mentioned, when he first became involved with the business, Joyce did not know Tim Horton at all. In fact, the man from Tatamagouche had never seen Horton play hockey, nor had he been to an NHL game. In that light, it is perhaps of interest to have a look at their first encounter. Joyce talks about it, and about Horton in his book on the business the two of them ran.

They met shortly after the departure of Horton's first partner. The company was then in a rather precarious state and Joyce knew something had to be done about it right away, so he called Horton and suggested they meet and try to iron out some of the problems. When the hockey player arrived, Joyce was somewhat surprised.

> [He expected a much taller man, and a] more imposing figure. He also wore very thick glasses, as his eyesight was terrible. He tried wearing contacts when he played hockey, but he kept losing them when he made contact with another player.
>
> He had the brush cut that was his trademark at the time, and he wore a suit.... His style was pretty standard for hockey players at that time — they were expected to adhere to a certain look in public.

Tim Hortons is a registered trademark of the TDL Marks Corp. Used with permission.

Tim Horton and Ron Joyce, around 1968.

Then, in perhaps the most commendable accolade possible, Joyce added: "Tim was enthusiastic and very trusting."[7] Such an observation was probably an indication of why the two men were able to work so well together in the years that followed.

After that initial meeting, and, in time, the most glaring problems were resolved, and gradually other stores were opened and the operating success became entrenched. After Tim Horton's death, Lori inherited a stake in the business, but eventually Joyce bought her out. There were then forty stores. He continued to grow the company, not only in Ontario, but across Canada and elsewhere. Aside from a few unsuccessful forays, the expansion has been largely successful. Today, of course, it is one of the most recognized businesses in Canada.

Over the years, there have been some aspects of the company that are widely known. One of them was the advent of something called the Timbit. These little products came out about 1976 and were quite popular from the outset. And to clarify one thing about them: despite their often being referred to as donut holes; they are not that at all. They are really just small, round donuts. Customers took to them immediately, and perhaps as Ron Joyce explained, "They felt that by eating Timbits, they were cutting back from eating donuts. In reality though, they would often consume six or seven … so they weren't really changing their eating habits at all."[8]

In the decades since the introduction of the little donut, its name has entered the lexicon of the land and has been adopted by other people for lots of things, but the small size and appeal remain. For example, during the month of August 2012, the summer Olympics were held in London, England. There were many athletic endeavours of course, but one of them was soccer.

As many sports fans will recall, the Canadian women's team defeated France 1–0, and won bronze medals for doing so. The player who scored the only goal in that game was a young woman named Diana Matheson. At five feet tall, she was the shortest athlete on the team, and as the heroine of the day became particularly endearing to people who knew her. One of those was Alison Meinert, who had employed the soccer star in prior years. After the victory over France, her former boss said affectionately of Matheson: "We called her Timbit — you could just pick her up and put her in your pocket."[9] No donut seller could have ever imagined such unorthodox advertising.

Another familiar coffee term was the so-called "Double-Double," or two sugar, two cream. If a customer ordered one of these, the counter person knew exactly what was wanted: no further explanation was needed. And even the general comments, "I'm going to Tim's," or "Meet me at Timmies," have become commonplace. In fact, it has been said that if you know what is intended by either remark, you are Canadian to the core.

Then there was and is a yearly contest the company runs called "Roll Up the Rim to Win." Perhaps no other purchase directive has the same familiarity for something that is a consumable product and a game of chance combined. This yearly contest is very familiar to company patrons wherever the chain sells its wares.

But the "Roll Up the Rim to Win" idea has a history in itself. Legend has it that it originated with a man named Ron Buist. He was a marketing guru for the company in 1986 who was looking for something that would be unique and profitable — and easy to understand. When he turned his attention to the take-out cups that were being used in the stores, he realized that each little container was an advertising item in itself. Unfortunately, it was also a readily identifiable trash item, too, if tossed aside heedlessly.

Photo courtesy of the author.

Always a welcome sign in any small town — the sometimes long-awaited arrival of a local Tim Hortons.

Buist studied the takeout cup more closely, and subsequently learned from the manufacturer that printing could be put on almost every part of its surface. That included on what was called its rim line, at the very top of the paper container. He was then informed that "anything printed there couldn't be seen when the rim was rolled down to form the cup's lip." From there, Buist told himself, "if you could not see the printing when the rim was rolled down, you could see it if the rim was rolled up."[10] This led directly to the placing of messages, such as "play again," or wording that in fact you were a winner of a coffee, a donut, or perhaps a box of Timbits. Later on, larger prizes were offered, as well.

As far as Buist was concerned, the idea met all the factors that were needed: his company "wanted to increase its sales; store owners who wanted to reward their customers, and customers who wanted an interesting, easy-to-win contest."[11] After all, the customer merely had to buy a coffee, and then turn up the lip in the top to see if they had won something. The idea took off, and remains to this day. Virtually anyone who has ever gone to one of the company stores knows of the idea, and may well have won from time to time.

Tim Hortons outlets are well known, and local operators have always been community-minded. For those reasons, the opening of a new store in almost any location is generally believed to be a worthwhile addition to its surroundings. In small towns in particular, the coming of a Tim-mies is good news. Then, once in operation, it easily fits into the fabric of its environment.

Originally, the company had an apostrophe in its name. Since then, "Tim Horton's Do-Nuts" has become "Tim Hortons," for simplicity's sake, but also because it now conforms to the language laws of the province of Quebec. There were some objections from the language purists when the punctuation mark was dropped, but that has largely disappeared.

And Tim Hortons is now an international behemoth compared to its humble roots in Hamilton almost five decades ago. As of July 2012, there were well over three thousand outlets in Canada; more than seven hundred in the United States; almost a dozen in the Persian Gulf area; and even a few in the U.K. and Ireland, operating out of the popular Spar convenience store chain. Now, the name on all those stores may refer to a phenomenally successful fast-food company and nothing more. Today, there are untold numbers of customers who have no idea who Tim Horton might have been. And while that may be disappointing to those who know hockey history, it would probably be okay with the Hockey Hall of Fame inductee himself. He simply wanted a business that would sustain him in retirement, and it certainly would have done that. If only he had been lucky enough to retire.

22

LESTER PEARSON
A Flag Unique in the World

Whether draped over the casket of a fallen soldier, held aloft by an Olympic medal-winner, or flying free from the Peace Tower in Ottawa, the red-and-white maple leaf flag is recognized by all Canadians. It is a proud symbol of our country, above the vastness of the prairies, the snows of the north, the border to the south. The flag means "Canada" wherever it flies: across our homeland, on our embassies abroad, on our military outposts where peace may be as deadly as war. This colourful piece of cloth means so much to so many, here and around the world.

For the most part, we love our flag, respect it, and cherish everything it represents. And that means a nation that is at peace; a nation that is like no other; a nation that means opportunity, civility, and the likelihood of a future filled with hope. Embodied in such lofty ideals is a country that is admired, envied, and truly blessed. It is our homeland, and few of us would trade it for anywhere else, now or in the future. Our flag reflects us and our Canada to the world.

But as flags go, ours is a relatively recent one. Canada became a nation in 1867, but it was almost a century before we got our own national flag. There were others over the years that were used for special occasions, or in particular areas of the country or in certain provinces, and some of those flags were beautiful, meaningful, and locally accepted. Certain ones were historical, battle-scarred, or emblematic of our early heritage, but they were sometimes divisive, derided, and not truly symbolic of this nation bounded by three oceans, a world power as a neighbour, and a geography so vast it is almost incomprehensible.

Neither of our two well-known earlier flags, the Red Ensign or the Fleur-de-lis, represented the country as a whole, nor did they reflect the multinational, multicultural, multicoloured face that this nation has become. There was nothing inherently wrong with either, but the Red Ensign in Chicoutimi was about as welcome as the Fleur-de-lis in Cranbrook. There was little connection to the community or to those who lived there. In fact, either could be regarded as provocative, out of place, and not at all representative of the locale. Both would reflect only a portion of the country as a whole, despite whatever historical or nationalistic connections they might have possessed individually.

But there was never complete agreement about what Canada's flag should be and even now, many years after the adoption of the maple leaf, it is not, nor ever will be universally accepted in Quebec. Anyone vaguely familiar with this country knows that. And we can either applaud that or lament it; it is a fact — and likely something that will never change. But insofar as the vast majority of the country, by population, geography, or racial component is concerned, the maple leaf flag embodies everything we believe our nation to be now, or will be in the future. And this flag is not yet a half-century old, but most Canadians can remember no other. The maple leaf may have been born in acrimony, divisiveness, and political partisanship, but it quickly became accepted and respected like no other institution in this nation. It seemed as if Canadians at large bore a sigh of relief when we — at last — had our own flag: a flag we saw as colourful and reflective of our country and ourselves. At last, we had our very own emblem to show to the world.

But getting there was far from easy.

When the great explorer Samuel de Champlain founded what is now Quebec City in 1608, what was called a French Merchant flag was hoisted over the fledgling settlement. A century and a half later, the divisiveness we see in this country today grew from what was really a small French-English skirmish on Abraham Martin's farm on the heights above the town. With the then raising of the British Union Jack in victory after the clash, the history of the New World started to change. Those of us who are even vaguely familiar with our past history know that. From then on, there would have been no hope at all there would ever be one flag for the new country that would be loved by everyone. But for over 350 years, whatever flag flew over our land was never *our* flag; it was Britain's, or a variant of the one at Westminster. In times of peace or war; in times of depression or prosperity; in times of geographical expansion; in times of population growth, cultural change, and political difference, we lived under a flag that was not our own. No wonder we needed a new flag. Fortunately one man in a position of power saw that and decided to do something about it.

His name was Lester Bowles Pearson, and he was the Nobel prize-winning prime minister of Canada from 1963 until 1968.

There had been others before Pearson who realized that Canada should have its own flag, but no one had the foresight, the stamina, or the political clout to do what Pearson did. Always it seemed, some group, faction, or self-interested segment of the population either did not like whatever was proposed, or they thought someone else was going to benefit inordinately. Always it seemed, at the end of the day nothing concrete happened.

Right after Confederation in 1867, there were many who felt that this new country should have its own flag. Several possibilities were suggested, argued over, and put aside. Canada's first governor general, a man named Viscount Charles Stanley Monck, had his own flag, a modified Union Jack. It was flown in Canada in his time in office here. Then as the years went on, the Red Ensign started to become popular. It was looked at more than once, sometimes as a temporary measure for a specific reason. By 1870, modified versions of the flag were flying over what were then new Canadian provinces. A little later, the British government gave official sanction for its use on our ships wherever they sailed. In 1925 Prime Minister Mackenzie King wanted a more distinctive flag for the country and he even set up a committee to look into the matter, but nothing came of it.

On the battlefields of the Second World War in particular, Canada's troops fought and died under the Red Ensign, and they developed a keen affinity for it. They regarded it as our nation's flag and felt that it should be forever enshrined as such. It was little wonder then that the Royal Canadian Legion would become the foremost proponent of the Red

Toronto Star and Archives Canada E002505448

Prime Minister Lester B. Pearson was the principal architect behind Canada's flag. Even though he was constrained by a minority government, he was still able to shepherd the necessary legislation through Parliament to give his nation its most lauded national symbol.

Ensign, as Lester Pearson soon learned when he decided to establish a new symbol to represent the nation. In a move some regarded as in-your-face bravado, he decided to tell the Legion first about his plans for a new flag. It is said that he even did so "without asking or telling his Cabinet colleagues,"[1] so sure was he in his intentions.

Because John Diefenbaker was always a staunch supporter of the Red Ensign, he bitterly opposed the adoption of the Maple Leaf flag.

The Legion was holding its annual convention in Winnipeg in 1964, and delegates conferred in various rooms, often below a portrait of Queen Elizabeth II, while the Red Ensign was in a position of honour nearby. The world of the Legion member was as it was meant to be — until Pearson arrived.

It was noted that he was wearing his military medals as he stepped to the rostrum in the main conference hall on May 17. Then, with his characteristic lisp and low-key speaking voice, he told the assembled delegates what he intended to do. They listened respectively at first, but then stood and booed. Before the prime minister had reached the airport to fly home, they were denouncing the plan to any reporter they could find.

At the time, John George Diefenbaker, Canada's former prime minister, whom Pearson had succeeded, was the leader of the official opposition in Ottawa. With little prompting, he took up the Legion cause. To him, the Red Ensign was Canada's flag and no one should declare otherwise. Canada was British, and must remain so. He trumpeted that view widely. "I want to make Canada all Canadian and all British," he said often, "the men who wish to change our flag should be denounced by every good Canadian."[2]

Obviously, Lester Pearson felt the heat, but he had no intention of backing down.

By the time he got back to Ottawa from Winnipeg, what came to be called the Great Flag Debate of 1964 had begun. It raged for several months, and many took up the cause, arguing for one side or the other. There were almost daily letters to the editor in the newspapers of the nation. Television and radio reports about the flag proposal dominated the airwaves. Op-ed pieces appeared and the talking heads of television entered the fray, said their piece, and repeated it ad infinitum. Ordinary Canadians sometimes took sides; sometimes not, depending on the company present, the location and type of bar, or the amount of alcohol consumed. Almost everywhere you went that summer, people talked of the flag, which flag, or why a new one should or should not be created. But sometimes you feared even mentioning the thing if you knew how friends felt.

I remember vividly that a colleague of mine used to refer to the Red Ensign during conversations that had nothing to do with the flag at all. He used to tell me how beautiful it looked flying above the snow in the far North, on post office flagpoles, even on customs offices at the Canadian border. I eventually stayed away from the subject completely when I talked to him. Many other people felt the same way that summer, in communities large and small. The flag question was something that could cause arguments, so that you avoided it in casual conversation, as you did politics and religion. In official Ottawa, of course, that was not the case at all.

As has been mentioned, over the years various committees had been struck, and each time those involved attempted to find a common ground in order to choose a flag that would be accepted by most Canadians. This had happened to the Mackenzie King committee of 1925 but it, like the others, was unsuccessful. Pearson knew the history of the problem, but resolved to accomplish something concrete this time. After all, "the election of the spring of 1963 had brought the Liberals back to power, but with a minority government. During the campaign, Pearson had promised that Canada would have a national flag within two years of his election, in plenty of time for the 1967 centennial of Confederation."[3]

The prime minister felt strongly about the need for a flag, and he expressed his sentiments in his memoirs. To him, "the flag was part of a deliberate design to strengthen national unity, to improve federal-provincial relations, to devise a more appropriate constitution, and to guard against the wrong kind of American penetration."[4] He felt that a new flag would aid in addressing these matters, but that it had to be selected as quickly as possible.

The ensuing Parliamentary debate on the flag issue was, as Pearson later admitted, "long, wearing, and at times depressing."[5] As always, the headline-grabbing assertions by John Diefenbaker seemed designed to throw a monkey wrench into the deliberations. When the Liberals introduced a motion for a flag, designed by Group of Seven painter A.Y. Jackson with two blue borders and three conjoined red maple leaves on a white background, Diefenbaker heaped unreserved scorn on the idea. "Surely Canada deserves something better than the symbol of three maple leaves,"[6] he thundered in Parliament. He also pointed out — erroneously — that Pearson's idea for the new flag bore "no relationship to Canada's past."[7] Then he coined a phrase that stuck when he referred to the suggested flag and its three maple leaves, as "Pearson's Pennant." The press and the public all picked up on the term, and in so doing caused Pearson more discomfort than was warranted.

Because there was no real progress in a final selection for any flag, Diefenbaker then demanded a national referendum on the matter. He seems to have had no precise plan for his request, but it could well have been an attempt to drag the issue out long enough so that his beloved Red Ensign might possibly be chosen, and the matter then could be closed. Had a referendum been held, any "new" flag likely would not have been ready for the centennial of the country, but to Diefenbaker, the Red Ensign would be. However, the referendum never happened because Pearson refused to hold it. Instead, he appointed another committee and told its members that they had six weeks to select a flag!

A principal member of the committee chosen by Pearson was a loyalist named John Matheson, a Liberal Member of Parliament from Leeds County, in Ontario. He was well aware of where the problems might lie and later wrote of his misgivings: "Choosing a flag was one thing and having it accepted with virtual unanimity was another."[8] He was right, and the committee contingent as well as most members of Parliament in that summer session had become quite riled up by the matter. In Matheson's view, "the invective, the rhetoric, the occasional flashes of eloquence in six months of debate in the Commons, all confirm how close this subject was to the bone, how near many members were at times to angry tears."[9] The committee knew they faced a massive task, but they moved quickly because they had to.

They held thirty-five rancorous meetings and most of them went nowhere. That was to no one's surprise. The committee looked at and discounted hundreds and hundreds of ideas, a vast majority of them submitted by members of the public — from all areas of the country and by people of all ages. There were grade-school drawings, university professors' submissions, and suggestions from individuals with almost every level of schooling in between. The task for the committee became more herculean as the time ticked down.

Then finally, there was a stroke of genius.

John Matheson made a move that could well have merited for him lasting acclaim from the nation. He knew a man in Kingston, Ontario, who was a historian at the Royal Military College. The individual's name was George F.G. Stanley, and, earlier on, he had given Matheson a possible flag design that was based on the flag of that school. Matheson liked it and showed the design to his fellow members in one of the final committee meetings. They liked it as well, and passed it along to Lester Pearson as their choice for the new flag. There was a subsequent parliamentary vote on the matter, and Doctor George Stanley's flag was chosen as the new Canadian flag. That red-and-white maple leaf is the one we have today.

As might be expected, John Diefenbaker was a disappointed and bitter man. He was a loyal Canadian, and he fought hard for the flag he loved, but in the end had to face up to the selection of others. Lester Pearson refused to gloat, but as history has shown us, he will always be remembered as the man who gave us the maple leaf flag. But John Matheson was especially pleased. In his thoughts about the flag, the committee work to select it, and the choice that was made, his love for the flag is obvious. "May the Maple Leaf fly for as long as the wind shall blow," he wrote at a later date. "May it be seen … as a signal from a kindly, caring, considerate people."[10]

Lester Pearson and John Diefenbaker quarrelled at length over what became known as the maple leaf flag. It flew for the first time over the Parliament buildings in Ottawa at exactly noon on February 15, 1965.

The final vote endorsing the new flag was held in the House of Commons in Ottawa on December 15, 1964, and Queen Elizabeth II proclaimed it a month later. Then the maple leaf flag of Canada was raised for the first time over Parliament Hill at exactly noon on February 15, 1965. It was widely accepted almost immediately.

NOTES

Chapter 1: Frederick Banting — Lifesaver for Millions

1. Harry Black, *Canadian Scientists and Inventors* (Markham: Pembroke Publishers, 1997), 14.
2. Frederick Banting, in Michael Bliss, *The Discovery of Insulin* (Toronto: University of Toronto Press, 2007), 50.
3. Bliss, 86.
4. Citation for Doctor Frederick Grant Banting, Canadian Medical Hall of Fame, London, Ontario.
5. Banting, in Bliss, 78.
6. Stephen Eaton Hume, *Banting: Hero, Healer, Artist* (Montreal: XYZ Publishing, 2001), 20.
7. Nancy Silcox, "The Art and Life of Dr. Frederick Banting," *Arabella* (Summer 2012), 262.
8. David Loch, *Arabella*, 272.
9. Frederick Banting, to A.Y. Jackson. "Frederick Banting (Sir Frederick Grant Banting)" Michel Bigue Art Galleries, Saint-Sauveur, Quebec.
10. Burton Feldman, *The Nobel Prize* (New York: Arcade Publishing, 2000), 276.

Chapter 2: Alexander Graham Bell — Pioneer Communicator

1. Carlotta Hacker, *Inventors* (Calgary: Weigl, 2000), 8.

2.   Nostbakken, Janis and Jack Humphrey, *The Canadian Inventors Book* (Toronto: Greey de Pencier, 1976), 100.
3.   Charlotte Gray, *Reluctant Genius* (Toronto: HarperCollins, 2006), 123.
4.   Alannah Hegedus and Kaitlin Rainey, *Bleeps and Blips to Rocket Ships* (Toronto: Tundra, 2001), 36.

Chapter 3: Mike Lazaridis — Genius Ahead of His Time

1.   Maxine Trottier, *Canadian Inventions* (Toronto: Scholastic, 2004), 43.
2.   John Micsinszki, in Rod McQueen, *BlackBerry* (Toronto: Key Porter, 2010), 23.
3.   *Ibid.*, 28–29.
4.   Mike Lazaridis, in McQueen, 31.
5.   *Ibid.*, 34.
6.   Mike Lazaridis, in Renata D'Aliesio, "Mike Lazaridis's Showcase Home Built to House His Large Ideas." *Globe and Mail*, February 4, 2012.
7.   Lazaridis, in McQueen, 45.
8.   McQueen, 65.
9.   *Ibid.*, 90.

Chapter 4: Henry Woodward and Mathew Evans — Two Friends Lighting the World

1.   Michael Shanks, "The Electric Light Bulb," *www.Archaeology.stanford.edu* (accessed 2012).
2.   "Electric Light, Henry Woodward and Mathew Evans," Science and Technology, *www.collectionscanada.gc.ca/innovations* (accessed 2012).
3.   Randall Stross, *The Wizard of Menlo Park* (New York: Crown, 2007), 3.
4.   Thomas Edison, in Paul Israel, *Edison, a Life of Invention* (New York: Wiley, 1998), 25.
5.   Thomas Edison, in Bruce Ricketts, "The First Electric Light Bulb," *Mysteries of Canada*, *www.mysteriesofcanada.com* (accessed 2012).
6.   Merriam-Webster's Collegiate Encyclopaedia, "Edison, Thomas Alva (1847–1931)," 511.
7.   Pat Brennan, "Great Minds Think Alike — Even When They Are On Holiday," *Toronto Star*, April 16, 2009.
8.   Stross, 2.

9. Edmund Morris, "Edison Illuminated," *The New York Times Book Review*," March 25, 2012.

Chapter 5: James Naismith — A Round Ball and a World Game

1. Frank Cosentino, *Almonte's Brothers of the Wind* (Burnstown, ON: General Store, 1996), 5.
2. Cosentino, 20.
3. Carlotta Hacker, *Inventors* (Calgary: Weigl, 2000), 38.
4. *Ibid.*
5. Harry Black, *Canadian Scientists and Inventors* (Markham: Pembroke Publishers, 1997), 84.
6. James Naismith, *Basketball: Its Origin and Development* (1941; reprint, Winnipeg, Bison Press, 1996), 53.
7. *Ibid.*, 57.
8. "Naismith Rules Raise Millions for Sports Charity," *Toronto Star*, December 11, 2010.

Chapter 6: Wilfred Bigelow — Heart Surgeon and Lifesaver

1. U.S. National Library of Medicine, Bethesda, Maryland.
2. W.G. Bigelow, *Cold Hearts* (Toronto: McClelland & Stewart, 1984), 11.
3. ____, 67.
4. ____, 90.
5. ____, 98.
6. W.G. Bigelow, M.D., J.C. Callaghan, M.D., and J.A. Hopps, "General Hypothermia for Experimental Intracardiac Surgery." Paper read before the American Surgical Association, Colorado Springs, Colorado, April 20, 1950.
7. Bigelow, 135.
8. Jen Clark, "Pacemaker," *The Beaver* (August/September 2009), 10.
9. Bigelow, 17.
10. *Canadian Encyclopedia*, Vol. 1, 215.
11. Dr. Wilfred Gordon Bigelow, Citation [in part], Canadian Medical Hall of Fame.
12. Wilfred Gordon Bigelow, O.C., Citation [in part], Order of Canada.

Chapter 8: Joseph-Armand Bombardier — The Man Who Conquered Winter

1. Larry Macdonald, *The Bombardier Story* (Toronto: John Wiley & Sons, 20010, 3.
2. Carlotta Hacker, *Inventors* (Calgary: Weigl, 2000), 19.
3. Carole Precious, *J. Armand Bombardier* (Toronto: Fitzhenry & Whiteside, 1984), 29.
4. *Ibid.*
5. Bev Spencer, *Made in Canada* (Markham, ON: Scholastic, 2003), 29.
6. Precious, 45.
7. Macdonald, 16.
8. *Canadian Encyclopedia*, Second edition, Vol. 1, 246.

Chapter 9: Norman Breakey — Wall-Painter's Dream

1. Beverley Tallon, "The Paint Roller," *The Beaver* (February/March 2009).
2. *The New York Times*, March 13, 1988.

Chapter 10: Charles Saunders — Nourishment for Canada and the World

1. Graham Chandler, "Selling the Prairie Good Life," *The Beaver* (August/September 2006), 26.
2. J.J. Brown, *The Inventors: Great Ideas in Canadian Enterprise* (Toronto: McClelland & Stewart, 1967), 107.
3. Elsie M. Pomery, *William Saunders and His Five Sons* (Toronto: The Ryerson Press, 1956) 74.
4. Daria Coneghan, *Encyclopedia of Saskatchewan* (Regina: University of Regina and Canadian Plains Research Center, 2007).
5. Pomeroy, 146.

Chapter 11: Jacques Plante — Facing the World's Fastest Game

1. Ted Denault, *Jacques Plante* (Toronto: McClelland & Stewart, 2009), 132.
2. Dan Diamond, ed., *Total Hockey* (Kansas City: Andrews McMeel Publishing, 1998), 1804.
3. Andy O'Brien, *The Jacques Plante Story* (Toronto: McGraw-Hill Ryerson, 1972), 14.
4. *Ibid.*, 15.

Chapter 12: Wilbur Franks — Safe Suiting for Flying

1. John Bryden, *Deadly Allies* (Toronto: McClelland & Stewart, 1989), 45.
2. Henry Head. "The Sense of Stability and Balance in the Air," in H. Milford, ed. *The Medical Problems of Flying* (London: Oxford University Press, 1920).
3. Dr. Wilbur Franks, "The Anti-G Suit," Banting Research Foundation Hall of Fame, University of Toronto.
4. Canadian Space Agency, *Canada's Aerospace Medicine Pioneers*, 6.
5. *Ibid.*, 7.
6. Aeromedical Handbook for Jet Passenger Ejection Seat Aircraft, 426 Training Squadron, CFB Trenton, Table 5.1.
7. Canadian Space Agency, 8.
8. J.J. Brown, *The Inventors* (Toronto: McClelland & Stewart, 1967), 112.
9. Ronald Wolf, "All Things Canadian — Wilbur Franks," *The Algoma News*, February 2011.

Chapter 13: Sandford Fleming — Time for All the World

1. T.D. Regehr, *The Canadian Encyclopedia*, Second Edition, Vol. 2 (Edmonton: Hurtig, 1988), 790–91.
2. Harry Black, *Canadian Scientists and Inventors* (Markham, ON: Pembroke, 1997), 58.
3. Janis Nostbakken and Jack Humphrey, *The Canadian Inventions Book* (Toronto: Greey de Pencier, 1976), 140.
4. Clark Blaise, *Time Lord* (Toronto: Alfred A. Knopf, 2000), 75.
5. Derek Hayes, *Canada, An Illustrated History* (Vancouver: Douglas and McIntyre, 2004), 45.
6. J.J. Brown, *The Inventors* (Toronto: McClelland & Stewart, 1967), 45.
7. Pierre Berton, *The National Dream* (Toronto: McClelland & Stewart, 1970), 46.

Chapter 14: Spar Aerospace — An Arm from Earth to Space

1. Lydia Dotto, *Canada in Space* (Toronto: Irwin Publishing, 1987), 202.
2. Canadian Space Agency, "The Development of a Legend."
3. *Ibid.*
4. John Melady, *Canadians in Space* (Toronto: Dundurn, 2009), 20.
5. Dotto, 210.

Chapter 15: Arthur Sicard — Making Winter Roads Passable

1. Mario Theriault, *Great Maritime Inventions* (Fredericton: Goose Lane Editions, 2001), 61.
2. Terry Frei and Adrian Dater, "Freak Snowblower Mishap Shelves Sakic," *The Denver Post*, December 11, 2008.
3. 88.9 Shine FM, 4510 MacLeod Trail South, Calgary, Alberta.
4. Susan Hughes, *Canada Invents* (Toronto: Maple Tree Press, 2002), 15.

Chapter 16: Harry Stevinson — Finding Plane Crashes and Saving Lives

1. John Melady, *Overtime, Overdue: The Bill Barilko Story* (Trenton, ON: City Print, 1988).
2. "Extraordinary Inventor," *Faculty of Engineering Magazine*, University of Alberta, Winter 2005, 1.
3. *Ibid.*
4. Crash Position Indicator, *http://en.wikipedia.org/wiki/Crash_position_indicator*.
5. "Saving Survivors by Finding Fallen Aircrafts," National Research Council Canada, 2012.
6. Sadiq Hasnain, "The Crash Position Indicator Aviation Study," IEEE Canada, April 1979, 2.
7. ____, 3.
8. Harry T. Stevinson, "Application for Crash Position Indicator for Aircraft." United States Patent Office, Washington, D.C., March 26, 1957.
9. *Ibid.*
10. *Ibid.*
11. "Extraordinary Investor," 5.

Chapter 17: Jim Floyd — The Airplane That Should Have Been

1. James Marsh, "Avro Jetliner," *Canadian Encyclopedia.* Vol. 1, 157.
2. Don Rogers, in Jim Floyd, *The Avro Canada C102 Jetliner* (Erin, ON: Boston Mills, 1986), 186.
3. "First Jet Liner Seen Here Flies from Toronto in Hour," *The New York Times*, April 19, 1950, 1.
4. Rogers, 191.

5. *Ibid.*
6. *Democrat and Chronicle* (Rochester, NY), January 12, 1951.
7. "What Happened to the Great American Aircraft Industry?" *Air Trails*, August, 1950.
8. J.J. Brown, *Ideas in Exile* (Toronto: McClelland & Stewart, 1967), 298.

Chapter 18: William Stephenson — Pictures without Distance

1. J.J. Brown, *The Inventors* (Toronto: McClelland & Stewart, 1967), 90.
2. H. Montgomery Hyde, *The Quiet Canadian* (London, UK: Hamish Hamilton, 1962), 7–8.
3. Bev Spencer, *Made in Canada* (Markham, ON: Scholastic Canada, 2003), 129.
4. Roald Dahl, quoted in William Stevenson, *A Man Called Intrepid* (New York: Harcourt Brace Jovanovich, 1976), 17.
5. David Stafford, *Camp X* (Toronto: Lester & Orpen Dennys, 1986), 13.

Chapter 19: Wallace Turnbull — All for the Way of Flight

1. Mary Turnbull, *Wallace Turnbull 1870–1954*. Canada Science and Technology Museum, Ottawa.
2. Keys to History, Wind Tunnel Description, McCord Museum, Montreal.
3. Harry Black, *Canadian Scientists and Inventors* (Markham, ON: Pembroke Publishers, 1997), 110.
4. *Ibid.*
5. J.J. Brown, *Ideas in Exile* (Toronto: McClelland & Stewart, 1967), 187.
6. *Ibid.*, 188.
7. Frank H. Ellis, *Canada's Flying Heritage* (Toronto: University of Toronto Press, 1954), 14.
8. Museum Reference, Heritage Branch, Province of New Brunswick, Saint John, 2008.

Chapter 20: Tom Patterson — Saving a Town with Culture

1. Carolynn Bart-Riedstra and Lutzen H. Riedstra, *Stratford: Its History and Its Festival* (Toronto: James Lorimer, 1999), 32.
2. *Ibid.*, 41.
3. The City of Stratford web site, *www.city.stratford.on.ca.*

4. L. Riedstra, "Stratford's Railway Industry," *www.acrossthebridgebandb. ca/images/hist_railway_industry.pdf*.

5. Tom Patterson, *First Stage: the Making of the Stratford Festival* (Toronto: McClelland & Stewart, 1987), 27.

6. *Ibid.*, 35.

7. Martin Knelman, *A Stratford Tempest* (Toronto: McClelland & Stewart, 1982), 7.

8. *Ibid.*

9. Patterson, 72.

10. Herbert Whittaker, *The Stratford Festival, 1953–1957* (Toronto: Clarke Irwin, 1958), xiv.

11. Jim Brown, author interview, August 24, 2012.

12. Stratford Festival Archives and History, "The Stratford Story," 2.

13. *Ibid.*

Chapter 21: Tim Horton and Ron Joyce — Double-Double Empire Men

1. Eli Gershkovitch, "Tale of Tim Horton's Last Ride," *Vancouver Sun*, September 30, 2005.

2. Timothy Feige, *Hockey's Greatest Tragedies* (London, UK: Arcturus, 2011), 103.

3. Kevin Shea, "One on One with Tim Horton," Hockey Hall of Fame, *www.hhof.com/htmlSpotlight/spot_oneononep197702.shtml*.

4. Ron Joyce, *Always Fresh* (Toronto: HarperCollins, 2006), 33.

5. Lori Horton, in Lori Horton and Tim Griggs, *In Loving Memory* (Toronto: ECW Press, 1997), 137.

6. *Ibid.*

7. Joyce, 43.

8. Joyce, 109.

9. Alison Meinert, in Wendy Gillis and Mark Zwolinski, "Oakville's Diana Matheson Never Made Limelight, Until She Won Canada Bronze," *Toronto Star*, August 10, 2012.

10. Ron Buist, *Tales from Under the Rim* (Fredericton: Goose Lane Editions, 2003), 97.

11. *Ibid.*, 95.

Chapter 22: Lester Pearson — A Flag Unique in the World

1. Norman Hillmer, "The Flag: Distinctly Our Own," *Canadian Encyclopedia*, 2012.

2. Peter C. Newman, "The Great Flag Debate," *The Distemper of Our Times* (Toronto: McClelland & Stewart, 1968), 254.

3. George F.G. Stanley, *The Story of Canada's Flag* (Toronto: McGraw-Hill Ryerson, 1965), 63.

4. Lester B. Pearson, *Mike* Vol. 3 (Toronto: University of Toronto Press), 270.

5. *Ibid.*, 273.

6. John Diefenbaker, *Canada; House of Commons Debates*, IV, 1964, 4330.

7. John Diefenbaker, *Montreal Gazette*, June 11, 1964.

8. John Ross Matheson, *Canada's Flag* (Belleville, ON: Mika Publishing, 1986), 126–27.

9. *Ibid.*, xiii.

10. Matheson, 252.

BIBLIOGRAPHY

Barbree, Jay. *Live From Cape Canaveral*. New York: HarperCollins, 2007.

Bart-Riedstra, Carolynn and Lutzen H. Riedstra. *Stratford: Its History and Its Festival*. Toronto: James Lorimer, 1999.

Berton, Pierre. *The National Dream*. Toronto: McClelland & Stewart, 1970.

Bigelow, W.G. *Cold Hearts*. Toronto: McClelland & Stewart, 1984.

Black, Harry. *Canadian Scientists and Inventors*. Markham: Pembroke Publishers, 1997.

Blaise, Clark. *Time Lord*. Toronto: Knopf, 2000.

Bliss, Michael. *Right Honourable Men*. Toronto: HarperCollins, 2004.

Bliss, Michael. *The Discovery of Insulin*. Toronto: University of Toronto Press, 2007.

Brown, J.J. *Ideas in Exile*. Toronto: McClelland & Stewart, 1967.

_____. *The Inventors*. Toronto: McClelland & Stewart, 1967.

Bryden, John. *Deadly Allies*. Toronto: McClelland & Stewart, 1989.

Buist, Ron. *Tales from Under the Rim*. Fredericton: Goose Lane Editions, 2003.

Cosentino, Frank. *Almonte's Brothers of the Wind*. Burnstown, ON: General Store, 1996.

Cruise, David and Alison Griffiths. *Lords of the Line*. Markham, ON: Viking, 1988.

Denault, Ted. *Jacques Plante*. Toronto: McClelland & Stewart, 2009.

Diefenbaker, John. *Canada: House of Commons Debates*. IV, 1964.

Diamond, Dan, ed. *Hockey Hall of Fame.* Toronto: Doubleday, 1996.

_____. *Total Hockey.* Kansas City: Andrews McMeel Publishing, 1998.

Dotto, Lydia. *Canada in Space.* Toronto: Irwin Publishing, 1987.

Ellis, Frank H. *Canada's Flying Heritage.* Toronto: University of Toronto Press, 1954.

English, John. *Citizen of the World: The Life of Pierre Elliott Trudeau.* Toronto: Knopf, 2006.

Feige, Timothy. *Hockey's Greatest Tragedies.* London, UK: Arcturus, 2011.

Feldman, Burton. *The Nobel Prize.* New York: Arcade, 2000.

Floyd, Jim. *The Avro Canada C102 Jetliner.* Erin, ON: Boston Mills, 1986.

Glover, Linda K. *National Geographic Encyclopaedia of Space.* Washington: N.D.

Gray, Charlotte. *Reluctant Genius.* Toronto: HarperCollins, 2006.

Hacker, Carlotta. *Inventors.* Calgary: Weigl, 2000.

Hayes, Derek. *Canada, An Illustrated History.* Vancouver: Douglas and McIntyre, 2004.

Hegedus, Alannah and Kaitlin Rainey. *Bleeps and Blips to Rocket Ships.* Toronto: Tundra, 2006.

Horton, Lori and Tim Griggs. *In Loving Memory: A Tribute to Tim Horton.* Toronto: ECW Press, 1997.

Hughes, Susan. *Canada Invents.* Toronto: Owl Books, 2002.

Hume, Stephen Eaton. *Banting: Hero, Healer, Artist.* Montreal: XYZ Publishing, 2001.

Hyde, H. Montgomery. *The Quiet Canadian.* London, UK: Hamish Hamilton, 1962.

Israel, Paul. *Edison, a Life of Invention.* New York: Wiley, 1998.

Joyce, Ron. *Always Fresh.* Toronto: HarperCollins, 2006.

Knelman, Martin. *A Stratford Tempest.* Toronto: McClelland & Stewart, 1982.

Macdonald, Larry. *The Bombardier Story.* Toronto: John Wiley & Sons, 2010.

Martin, Lawrence. *The Presidents and the Prime Ministers.* Toronto: Doubleday, 1982.

Matheson, John Ross. *Canada's Flag.* Belleville. ON: Mika, 1986.

Mayer, Roy. *Scientific Canadian.* Vancouver: Raincoast, 1999.

McQueen, Rod. *BlackBerry.* Toronto: Key Porter, 2010.

Melady, John. *Canadians in Space: The Forever Frontier.* Toronto: Dundurn, 2009.

_____. *Overtime, Overdue: The Bill Barilko Story.* Trenton, ON: City Print, 1988.

_____. *Pearson's Prize.* Toronto: Dundurn, 2006.

Milford H., ed. *The Medical Problems of Flying.* London: Oxford University Press, 1920.

Mulberry, Larry. *Aviation in Canada.* Toronto: McGraw-Hill Ryerson, 1979.

Naismith, James. *Basketball: Its Origin and Development.* 1941. Reprint, Winnipeg: Bison Books, 1996.

Newman, Peter C. *The Distemper of Our Times.* Toronto: McClelland & Stewart, 1968.

Nostbakken, Janis and Jack Humphrey. *The Canadian Inventions Book.* Toronto: Greey de Pencier, 1976.

O'Brien, Andy. *The Jacques Plante Story.* Toronto: McGraw-Hill Ryerson, 1972.

Pain, Richard and Christopher Plummer. *Stratford.* Erin, ON: Boston Mills, 1998.

Patterson, Tom. *First Stage: The Making of the Stratford Festival.* Toronto: McClelland & Stewart, 1987.

Pearson, Lester B. *Mike, Vol. 3.* Toronto: University of Toronto Press, 1975.

Peden, Murray. *Fall of an Arrow.* Stittsville, ON: Canada's Wings, 1978.

Phinney, Sandra. *Risk Takers and Innovators.* Canmore, AB: Altitude Publishing, 2004.

Platt, Richard. *Eureka!* Boston: Kingfisher, 2003.

Plummer, Christopher. *In Spite of Myself.* Toronto: Knopf, 2008.

Pomery, Elsie M. *William Saunders and His Five Sons.* Toronto: The Ryerson Press, 1956.

Precious, Carole. *J. Armand Bombardier.* Toronto: Fitzhenry & Whiteside, 1984.

Shea, Kevin. "One on One with Tim Horton." Hockey Hall of Fame, *www.hhof.com/htmlSpotlight/spot_oneononep197702.shtml.*

Spencer, Bev. *Made in Canada.* Markham, ON: Scholastic, 2003.

Stafford, David. *Camp X.* Toronto: Lester & Orpen Dennys, 1985.

Stanley, George F.G. *The Story of Canada's Flag.* Toronto: McGraw-Hill Ryerson, 1965.

Stevenson, William. *A Man Called Intrepid*. New York; Harcourt Brace Jovanovich, 1976.

Stevenson, William. *Intrepid's Last Case*. New York: Villard, 1983.

Stewart, Greig. *Shutting Down the National Dream*. Toronto: McGraw-Hill Ryerson, 1988.

Stross, Randall. *The Wizard of Menlo Park*. New York: Crown, 2007.

Theriault, Mario. *Great Maritime Inventions, 1833–1950*. Fredericton: Goose Lane Editions, 2001.

Trottier, Maxine. *Canadian Inventors*. Markham, ON: Scholastic, 2004.

Verstraete, Larry. *Whose Bright Idea Was It?* Markham, ON: Scholastic, 1997.

Whittaker, Herbert, Fwd. *The Stratford Festival, 1953–1957*. Toronto: Clarke, Irwin, 1958.

Wolfe, Tom. *The Right Stuff*. New York: Bantam, 1979.

ALSO BY JOHN MELADY

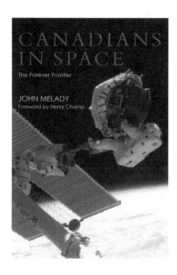

Canadians in Space
The Forever Frontier
978-1550029406
$27.99

Marc Garneau became Canada's first astronaut when he rocketed into space from a launch pad at Cape Canaveral, Florida. In doing so, he became a national hero. Seven of his fellow citizens followed in his footsteps, many more than once. Julie Payette, a young mother and adventurer from Montreal, was the first Canadian woman to visit the International Space Station. Chris Hadfield, a former fighter pilot from Ontario, was the first Canadian to do a spacewalk, while Saskatchewan-born Dr. Dave Williams performed surgery on test animals while his shuttle sped around the globe. This book was written as a twenty-fifth anniversary tribute to these brave men and women who defied tremendous odds, risked their lives, and soared from Earth on sheets of flame. By leaving the only planet known to be habitable, they became true explorers in an ever-expanding universe we will never completely know.

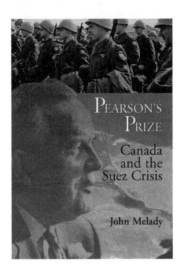

Pearson's Prize
Canada and the Suez Crisis
978-1550026115
$30.00

In the fall of 1956, the world was on the brink of war. Egyptian President Gamel Nasser nationalized the Suez Canal, and Britain, France, and Israel attacked him. Russia supported Nasser, and Soviet Premier Khrushchev threatened nuclear holocaust if the United States became militarily involved. Soon, the matter became a major problem for the United Nations. Fortunately, because of the efforts of Lester Pearson, then Canada's minister of external affairs, the crisis was defused. Pearson proposed a U.N. peacekeeping force be sent to Egypt to separate the warring factions there and keep the peace. Because his idea was adopted, Pearson helped save the world from war. For his outstanding statesmanship, Pearson won the Nobel Prize for Peace, the only Canadian ever to do so. This book, written to commemorate the fiftieth anniversary of the event, is about the Suez and about Pearson's work during a tension-filled time in the twentieth century.

Visit us at
Definingcanada.ca
@dundurnpress
Facebook.com/dundurnpress